Revised edition

New Wave Mental Maths E

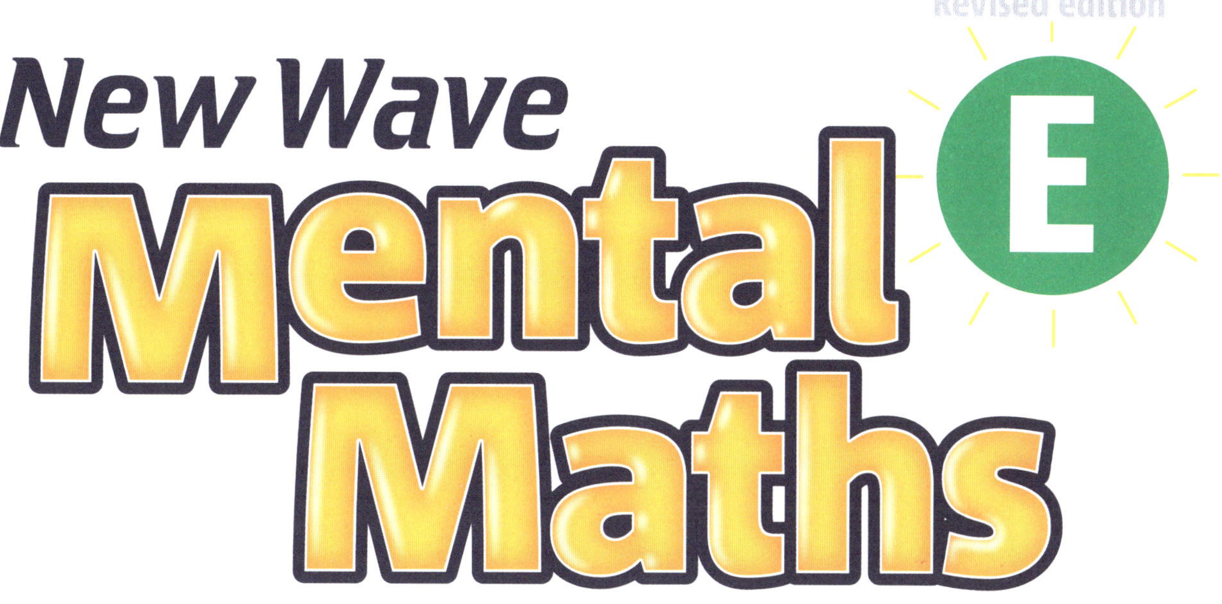

This book belongs to

....

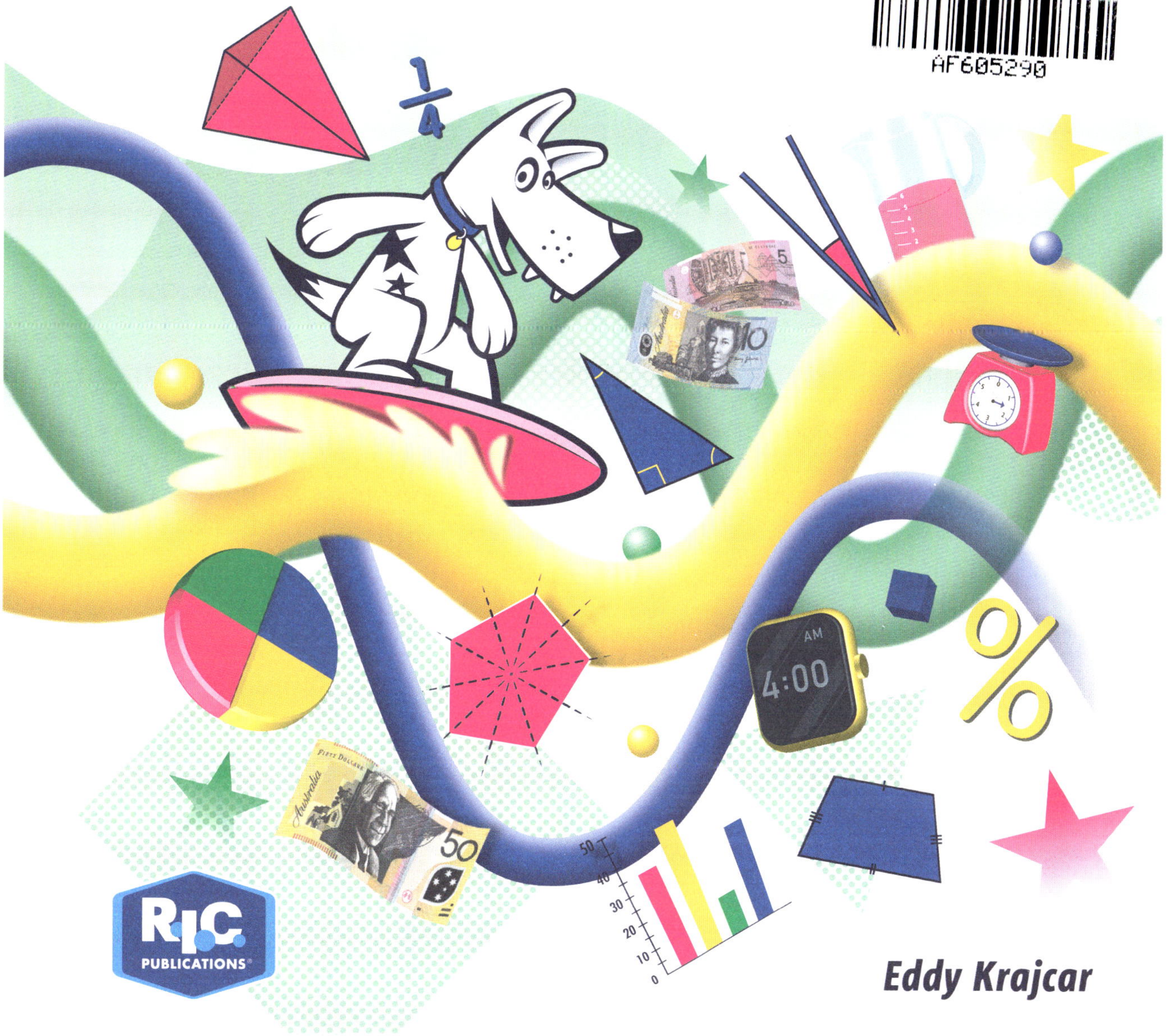

R.I.C. PUBLICATIONS®

Eddy Krajcar

Published by R.I.C. Publications®
PO Box 332, Greenwood
Western Australia 6924
+61 8 9240 9888
www.ricpublications.com.au
mail@ricpublications.com.au

First published 1999
Rewritten and reprinted 2011
Revised and reprinted 2012, 2017
Rewritten and reprinted 2024
Reprinted 2025

RIC-8618

ISBN 978-1-923005-52-5

R.I.C. Publications® acknowledges the Wadjak people of the Nyoongar Nation as the Traditional Custodians of the land on which our Western Australian office is based. We acknowledge the Traditional Custodians of Country throughout Australia and pay our respects to Elders past and present. R.I.C. Publications® recognises the role of First Nations Elders as Australia's first educators.

This book features several artworks illustrated and approved for appropriateness and accuracy by Melinda Brown, Spirit Dreaming, Ngunnawal Country.

Foreword

New Wave Mental Maths is a best-selling series written to consolidate students' mathematical understanding through comprehensive and structured daily mental maths practice. The series supports learning for curricula across Australia, and has been revised to meet the proficiencies of the Australian Curriculum Version 9.0, covering Understanding, Fluency, Problem-solving, and Reasoning.

The seven workbooks are designed to ensure students:

- practise mental calculation concepts and skills across all strands
- consolidate their understanding of mathematical topics and vocabulary
- reinforce their learning through self-guided revision of pre-taught concepts.

New to this edition is the 'Weekly Focus', which highlights a key topic covered throughout the week's learning, offering an ideal opportunity for class discussion. Additionally, the updated 'Problem-solving' column provides students the space to apply critical thinking and inquiry skills to in-depth problems.

The 'Friday Review' consolidates the week's learning with colour-coded questions representing the curriculum strands: blue for Number, purple for Algebra, green for Measurement, orange for Space, red for Statistics, and black for Probability. Designed to be flexible, educators may choose to use the 'Friday Review' as a pre- or post-assessment, or to allow students to self-assess and monitor their own progress.

The 'Maths Facts', located at the back of each workbook, support student learning through visual representation of mathematical concepts.

The goal of this series is to encourage students to practise mental calculations every day, supporting their future in a world that is full of maths.

Contents

Week 1

Monday

1. What is the time? 

   ________ pm

2. 3 × 3 × 2 = ______

3. Write the number shown on the abacus.

   ____

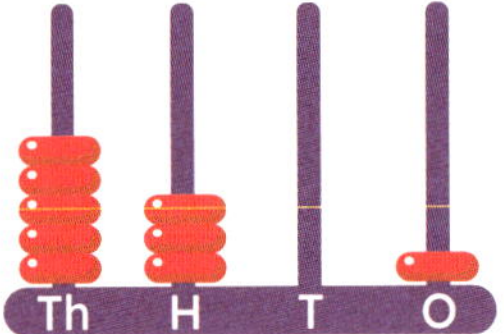

4. 4 × 5 = ______

5. 20 ÷ 4 = ______

6. Write $\frac{7}{100}$ as a decimal. ______

7. This is a 

   ________________.

8. Does sunset occur during am or pm?

   ____

9. 3 × 6 = 6 + 6 + 6 = ______

10. Round 4766 to the nearest thousand.

    ____

11. How many days are in a fortnight?

    ______

12. Together, Alex and Mimi ate 12 pieces of chocolate. Alex ate twice as much as Mimi. How many pieces did Alex eat? 

    ______

13. How many odd numbers are there between 1 and 10?

    ______

14. 13 – 9 = ______

15. 2.9 > 0.29 true ☐ false ☐

Tuesday

1. What is the time? 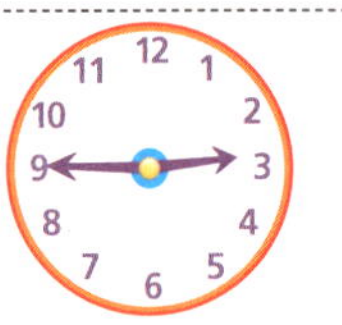

   ________ am/pm

2. 7 × 7 = ______

3. 800 + 700 = ______

4. 12 ÷ 3 = ______

5. 20 – 6 = ______

6. How far is it to Fords from the sign if Moppa is 8km further away than Fords? 

   ____

7. This polygon is known as a 

   ________________.

8. What is the date of the extra day in a leap year?

   ____

9. 20 × 6 = 120, 19 × 6 = 114,

   18 × 6 = ______

10. Which is heavier, 1kg or 700g? ______

11. 15 – 7 = ______

12. Which equation (number sentence) is equal to 9 × 7?

    60 + 3 = 63 ☐ 70 – 9 = 61 ☐
    50 + 40 = 90 ☐ 90 + 7 = 97 ☐

13. Complete the multiples of 3.

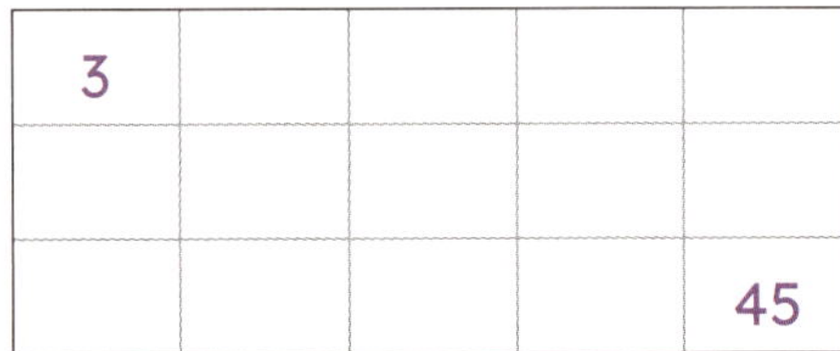

3				
				45

14. $\frac{1}{4} + \frac{1}{4} =$ ______

15. As a fraction, write the probability of landing on a 4? 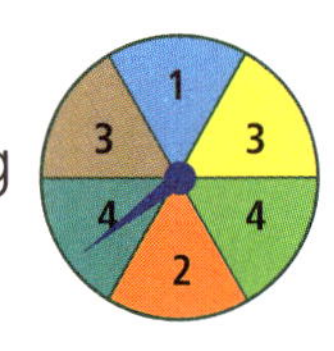

    ____

Wednesday

1. What is the time in the evening?

______

2. 9 × 9 = ______

3. 9 ÷ 3 = ______

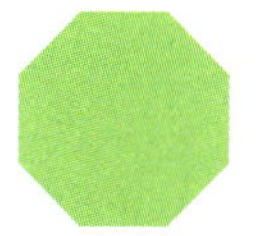

4. This is an ______________.

5. Write twelve thousand, eight hundred and one as a numeral.

______

6. 2, 4, 8, 16, ______

7. How many even numbers are there from 10 to 20?

______

8. If yesterday was Saturday, what day will tomorrow be?

______

9. If you travel west and turn right at the second street, which street are you in?

______

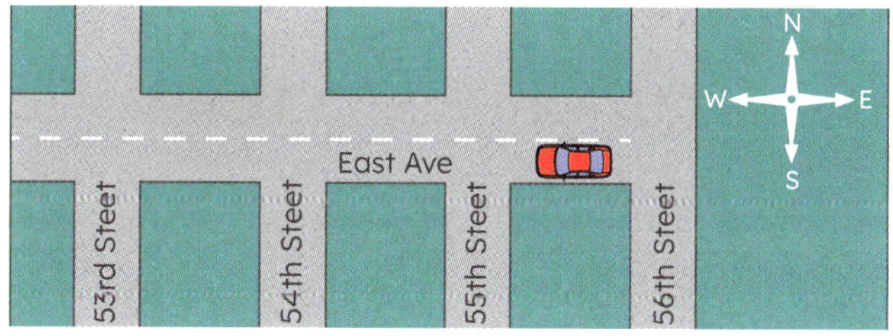

10. One decade = ______ years

11. 1100 – 300 = ______

12. 60 ☐ 5 = 12

13. Double 175. ______

14.

= $______

15. What is the perimeter of a regular hexagon with 6cm sides?

______

Thursday

Week 1

1. What is the time in the morning?

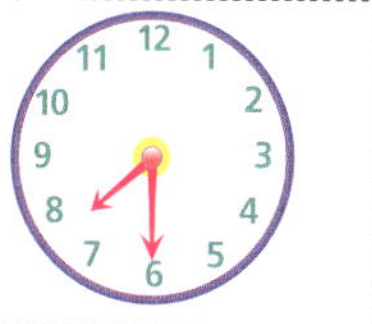

______

2. 110 – 50 = ______

3. This shape is an

______________.

4. 6 × 10 = 60, 6 × 100 = 600, 6 × 1000 = 6000, 6 × 10 000 = ______

5. How many weeks are in one year? ______

6. 15 ÷ 3 = ______

7. $\frac{4}{9} + \frac{2}{9} =$ ______

8. 1024 – 25 = ______

9. One century = ______ years

10. If the sun is in the east and is low to the horizon, is it morning or afternoon?

______

11. Write the number before 510. ______

12. 4 + 7 = ______

13. (a) Which year group is the zoo most popular with?

______

(b) To promote the zoo, which year groups would you target?

______

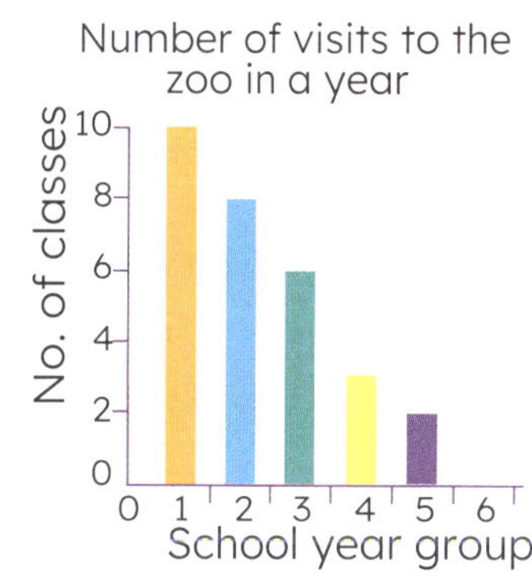

14. 1.1 > 1.04 true ☐ false ☐

15. Write $\frac{2}{100}$ as a decimal. ______

978-1-923005-52-5

Week 1

Problem-solving

The hands on the clock face show a right angle with both hands somewhere between the 7 and 11.

Let's find out what time it could be. Show your answers as 12-hour and 24-hour time.

Read the question again. **Think** about the information. Underline the important words.

Tick the strategy you will use to work out the answer:

- estimate and check ☐
- look for patterns ☐
- draw a diagram or picture ☐
- construct a table or graph ☐
- use materials ☐
- act it out ☐
- work backwards ☐
- something else. ☐

Solve it:

Reflect on the question and answer.

Check it. Circle another strategy on the list to work out the answer.

Show it:

Friday Review

1. $11 - 5 =$ ______
2. $12 \div 3 =$ ______
3. 10, 20, 40, ______, 160
4. $\frac{3}{9} + \frac{5}{9} =$ ______
5. On Tuesday's spinner, what is the probability of landing on an even number? ______
6. Write the number shown on the abacus. ______

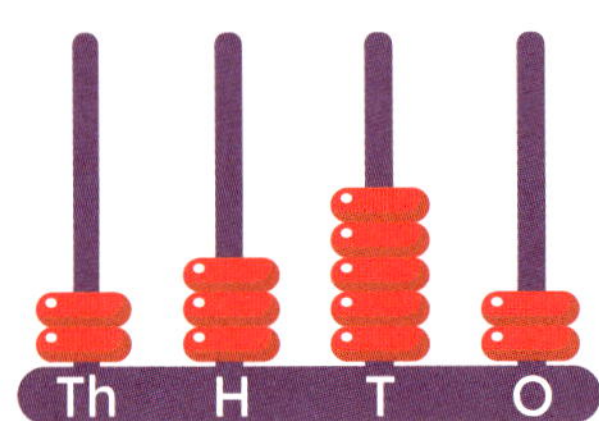

7. Together, Natasha and Sonja ate nine pieces of chocolate. If Sonja ate twice as much as Natasha, how many pieces did Sonja eat? ______
8. $8 < 10$ true ☐ false ☐
9. $20 \times 7 = 140$

   $19 \times 7 = 133$

   $18 \times 7 =$ ______
10. Double 175. ______
11. Write eleven thousand, one hundred and ten as a numeral. ______
12. In Wednesday's street map, if you travel east and turn left, which street are you on? ______
13. Write $\frac{4}{100}$ as a decimal. ______
14. From 1 Jan to 31 Dec is ______ days or one ______.
15. What is the time? ______

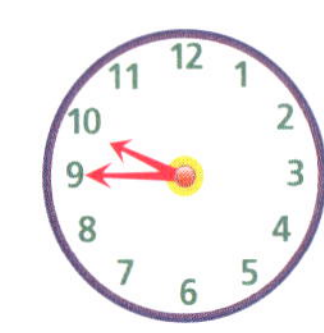

16. This is a ______.

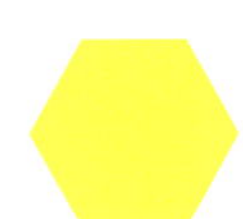

17. Which is heavier, 2kg or 400g? ______
18. The difference between the most to least favourite beach is ______ students.

Year 5 favourite beach for selfies

Favourite beach	Number of students
Lucky Bay	10
Trigg Beach	12
Eagle Bay	6

(Axis: Number of students 0, 4, 8, 12, 16; Favourite beach)

Sp

P

Monday

1. Measure the length of $\overline{AB}$ in cm.

   ______cm

   A ——————————— B

2. 5 × 9 = ______

3. This is a 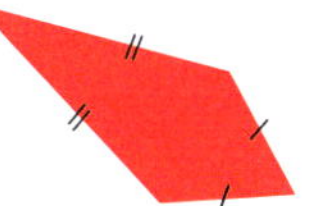

   ________________.

4. 1 minute = ______ seconds

5. 160, 80, ______, 20, 10

6. How far is it from the sign to Two Rocks if Yanchep is 10km further away than Two Rocks?

   ____

7. 24 ÷ 4 = ______

8. 3 × 7 = 7 + 7 + 7 = ______

9. Draw another line of symmetry.

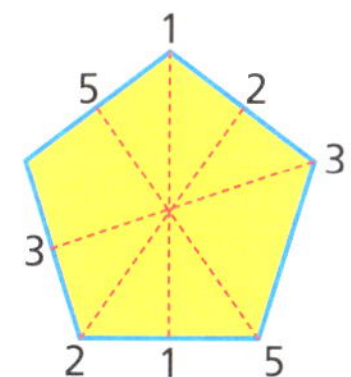

10. 5 × 5 × 20 = ______

11. 70 + 7 + 4 = ______

12. 120 – 70 = ______

13. Round 35 578 to the nearest ten thousand.

    ____

14. Olivia had  and spent

     +  +  + 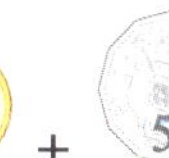.

    What amount of money was left? ______

15. $\frac{3}{5} + \frac{1}{5} =$ ______

Tuesday

Week 2

1. Measure the length of $\overline{AB}$ in cm.

   ______cm

   A ——————— B

2. What is the time? ______

3. Double 145. ______

4. If 1 hour is 60 minutes, and 3 hours is 3 × 60 = 180 minutes, then

   5 hours is 5 × ______ = ______ minutes.

5. 50 ÷ 5 = ______

6. 28 ☐ 4 = 7

7. Share 30 balloons into groups of six. (Write as a number sentence.) How many balloons per group?

   ____

8. This is a 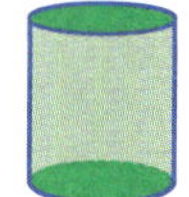

   ________________.

9. What is the cost of 2kg of bananas at $1.50 per kg?

   ____

10. Round 15 798 to the nearest thousand.

    ____

11. 60 + 6 + 7 = ______

12. If 1783 is 17 hundred and 83, then

    2495 is ______ hundred and ______.

13. 1cm = ______mm

14. 23 × 8 = (20 × 8) + (3 × 8)

    = ______ + ______

    = 184

15. odd – even = ______

Wednesday

1. Measure the length of $\overline{XY}$ in cm.

   ______cm

   X ———————— Y

2. What is the time? ________

3. Will 21 ÷ 3 equal a number greater than 10 or less than 10?

   ____

4. Write forty thousand and four as a numeral.

   ____

5. 1050, ________, 750, 600, 450

6. 5734 – 734 = ________

7. How many hours are in a day? ______

8. $5.00 – $1.90 = ________

9. This is an irregular ________________.

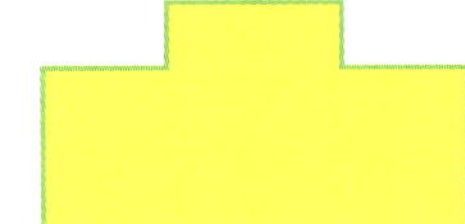

10. 3 × 8 = 8 + 8 + 8 = ______

11. On holidays you stop and read this road sign. What do the numbers represent?

Yarram	72
Welshpool	101
Naarm / Melbourne	291

   population ☐

   distance in miles ☐

   distance in kilometres ☐

12. 1L = ________mL

13. $\frac{3}{8} + \frac{4}{8} =$ ________

14. How many tens are there in 640? ______

15. Add 100 to 3980. ________

Thursday

1. Measure the length of $\overline{EF}$ in cm.

   ______cm

   E ———— F

2. Two equilateral triangles are joined at A and B. They make a:

   pentagon. ☐

   rhombus. ☐

   rectangle. ☐

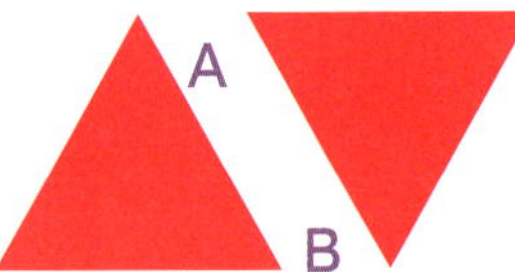

3. Write the fractions in ascending order.

   $\frac{1}{5}$ $\frac{1}{2}$ $\frac{1}{4}$ $\frac{1}{10}$ $\frac{1}{6}$

   ______ ______ ______ ______ ______

4. Double 275. ________

5. 45 ÷ 9 = ______

6. 1t = ________kg

7. 1km = ________m

8.   

   = $________

9. Write $\frac{4}{100}$ as a decimal. ________

10. In Wednesday question 11, what are the distances between:

    (a) Welshpool and Naarm / Melbourne? ________

    (b) Yarram and Naarm / Melbourne? ________

11. Read the pie graph and calculate the number of students that ate pasta.

    ______

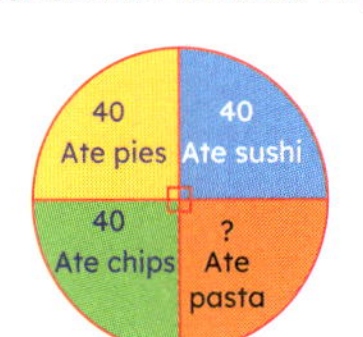

12. 13 – 8 = ______

13. odd + odd = ________

14. What is the perimeter of an equilateral triangle with 8cm sides? ________

15. What is the cost of 3kg of grapes at $2.50 per kg? ________

Problem-solving

Jade is training for a race. The track is 800 metres long, and they need to stop for 20mL of water every 10 000cms.

Let's find out what the minimum capacity of a drink bottle Jade needs is.

Read the question again. **Think** about the information. Underline the important words.

Tick the strategy you will use to work out the answer:

- estimate and check ☐
- look for patterns ☐
- draw a diagram or picture ☐
- construct a table or graph ☐
- use materials ☐
- act it out ☐
- work backwards ☐
- something else. ☐

Solve it:

Reflect on the question and answer.

Check it. Circle another strategy on the list to work out the answer.

Show it:

Friday Review

1. 600, 450, 300, ______

2. Write the fractions in ascending order.

   $\frac{1}{8}$ $\frac{1}{6}$ $\frac{1}{4}$ $\frac{1}{2}$ $\frac{1}{5}$

   ______ ______ ______ ______ ______

3. 2100 – 900

   = ______

4. 22 × 8

   = (20 × 8) + (2 × 8)

   = ______ + ______

   = ______

5. How far is it from the sign to Geelong if Drysdale is 20km further away than Geelong?

   ______ km

6. Add 100 to 2970.

   ______

7. Kieran had $10 and spent $5 + $1 + 50c. How much money was left?

   ______

8. $\frac{3}{9} + \frac{4}{9} =$ ______

9. Round 12 345 to the nearest thousand.

   ______

10. 8 + 8 + 8 = ______

11. Does 24 ÷ 3 equal an amount < 10 or > 10?

    ______

12. odd + even

    = ______

13. If Marissa and Angie together rode 60km in one week and Marissa rode twice as far as Angie, how far did Marissa ride?

    ______

14. What is the time?

    ______

15. 1kg = ______ g

16. Draw another line of symmetry.

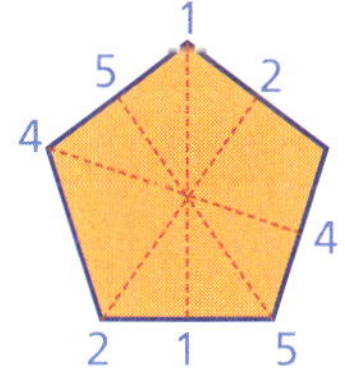

17. Name this shape.

    ______

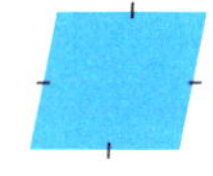

18. Measure $\overline{XY}$.

    ______ cm

    X ———— Y

Week 3

Monday

1. Rotate a $\frac{1}{4}$ turn clockwise.

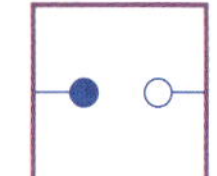 

2. Round 1433 to the nearest ten.

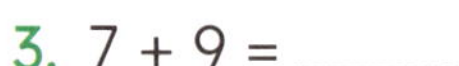

3. $7 + 9 =$ ______

4. $200 \times 3 = 600$, $201 \times 3 = 603$, $202 \times 3 = 606$, $203 \times 3 =$ ______

5. This solid 3D object is a ______ ______.

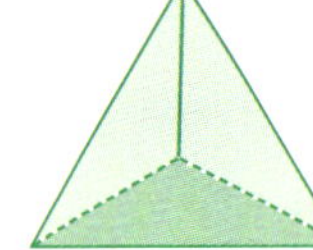

6. 1 whole and $\frac{3}{10} = 1.3$, so 1 whole and $\frac{6}{10}$ = ______.

7. Write the fractions in descending order.

$\frac{1}{3}$ $\frac{1}{5}$ $\frac{1}{10}$ $\frac{1}{4}$

______ ______ ______ ______

8. (a) $4 \times 9 =$ ______
   (b) $6 \times 9 =$ ______

9. Is $\frac{4}{5}$ or $\frac{2}{5}$ the larger fraction? ______

10. What is the perimeter of this rectangle? ______

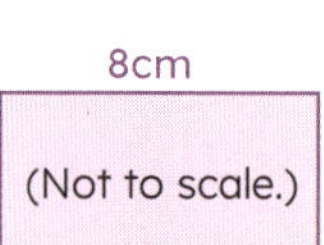

11. Which letter has no line of symmetry?

N E H M ______

12. What date is four months prior to 1 February?

______

13. What is the cost of buying 5kg of cheese at $9.00 per kg?

______

14. $11 + 11 + 11 =$ ______

15. Draw hands to show 4:55.

Tuesday

1. Rotate a $\frac{3}{4}$ turn clockwise.

2. $\frac{2}{3} + \frac{1}{3} =$ ______

3. $80 + 9 + 700 + 3000 =$ ______

4. Complete the pattern.

45, 50, 60, 75, ______, 120

5. How many more students liked Lucky Bay than Eagle Bay?

______

6. What is the length of A? ______m

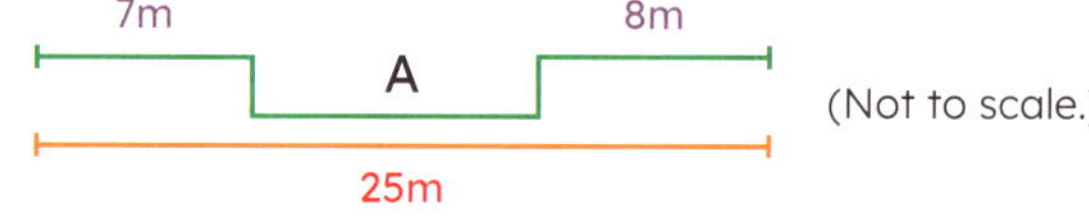

7. $34 \times 8 = (30 \times 8) + (4 \times 8)$
   = ______ + ______
   = ______

8. $1 + \frac{8}{10} = 1.8$, so $1 + \frac{4}{10} =$ ______.

9. Round 784 to the nearest 100. ______

10. 4.3cm = 43mm, so 8.5cm = ______mm.

11. $5 \times 8 =$ ______

12. $5.00 – $1.20 = ______

13. Round 345 034 to the nearest ten thousand.

______

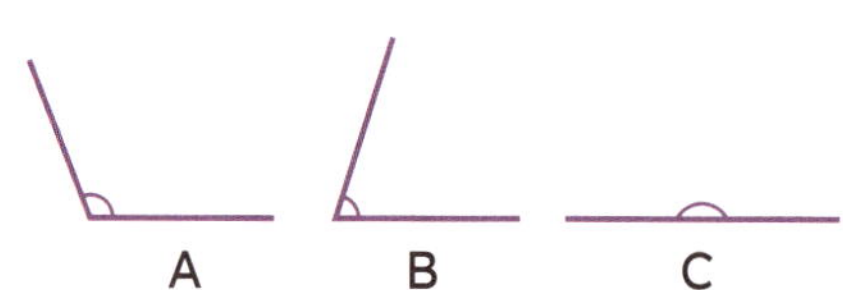

14. Which angle is 180°? ______

15. Which angle is obtuse? ______

Wednesday

1. Rotate a $\frac{1}{2}$ turn anticlockwise.

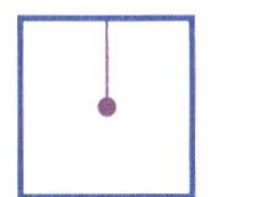

2. Draw hands to show 12:10.

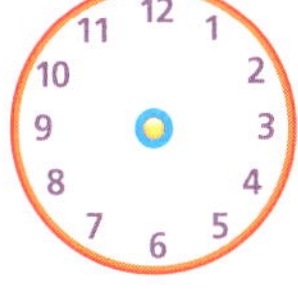

3. 3.5kg = 3500g, so 4.3kg = ________ g.

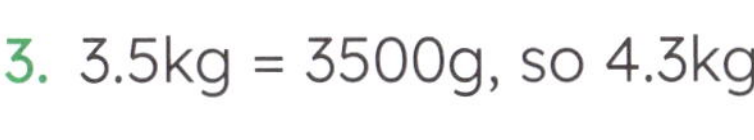

4. Halve 350. ________
5. Round 507 908 to the nearest ten thousand. ________
6. 2500 + ________ = 20 000
7. \$10.00 – \$4.10 = ________
8. Calculate the perimeter of a square with 7cm sides. ________
9. Which letter is symmetrical? ________

   I L Q S

10. Double 185. ________
11. How many odd numbers are there from 15 to 24? ________
12. 15 = 4 × 3 true ☐ false ☐
13. If you are cycling on 11th Road heading away from Main Road, what is your direction of travel?

   ____

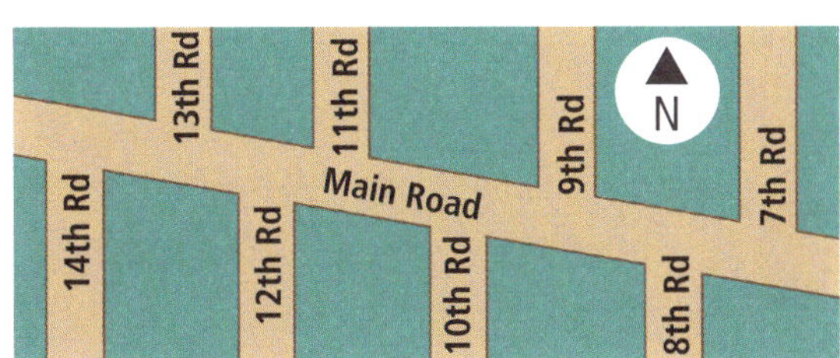

14. How many kilograms of salt would you receive for \$18.00 if the salt is \$2.00 per 10kg? ________
15. How many quarters make up a whole? ________

Thursday

1. Rotate a $\frac{1}{4}$ turn anticlockwise.

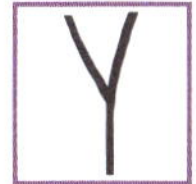

2. (a) 3 × 7 = ________ (b) 6 × 7 = ________
3. 4 + 5000 + 60 + 900 = ________
4.

   = \$________

5. Is $\frac{1}{4}$ or $\frac{2}{3}$ the smaller fraction? ________
6. Share \$10.00 equally among four children. ________ each
7. Double 495. ________
8. Reflect.

9. 1450, 1150, 850, ________
10. 6 × 6 = ________
11. This solid 3D object is a ________.

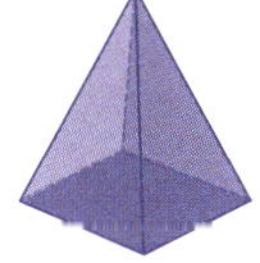

12. How far is it from the sign to Geelong if Torquay is 20km further away than Geelong? ________

13. 27 = 9 × 4 true ☐ false ☐
14. How many times will ten go into three hundred? ________
15. 30 × 9 = 270   29 × 9 = 261

    28 × 9 = ________   27 × 9 = 243

Week 3

Week 3

Problem-solving

Theo put their watch on upside down. Telling the time was tricky because the numbers had worn off.

Theo's view

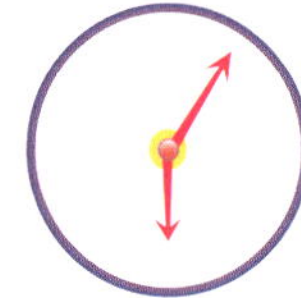

Let's find out the actual time.

Read the question again. **Think** about the information. Underline the important words.

Tick the strategy you will use to work out the answer:

- estimate and check ☐
- look for patterns ☐
- draw a diagram or picture ☐
- construct a table or graph ☐
- use materials ☐
- act it out ☐
- work backwards ☐
- something else. ☐

Solve it:

Reflect on the question and answer.

Check it. Circle another strategy on the list to work out the answer.

Show it:

Friday Review

1. 4 × 4 = ________
2. 60 + 4 + 300 + 2000 = ________
3. What is the chance of picking a blue marble from a jar of three blue and six red marbles? ________
4. Is $\frac{3}{5}$ or $\frac{1}{2}$ the smaller fraction? ________
5. Round 1737 to the nearest ten. ________
6. 1000, 800, 1200, 1000, 1400, ________
7. 2078 – 100 = ________
8. How many kilograms of lemons would you receive if you spent $25 and lemons cost $5 per 5kg? ________
9. 202 × 4 = 808
   203 × 4 = ________
10. How far is it from the sign to Manly if Manly is 20km further away than Warrane / Sydney? ________

11. In Tuesday's graph, how many students preferred Trigg Beach? ________
12. Rotate a $\frac{3}{4}$ turn anticlockwise.

13. How many days are there in March? ________
14. Draw an X at the southern end of 12th Rd.

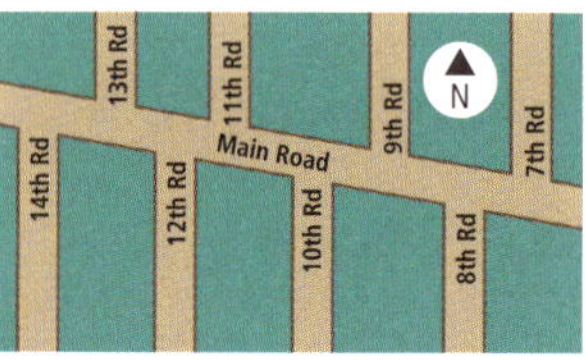

15. Name this 3D solid object. ________

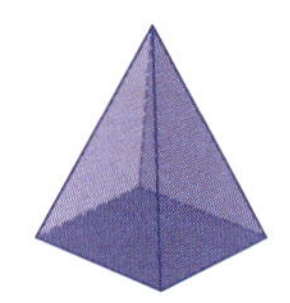

16. 2.7kg = 2700g, so 3.2kg = ________g.
17. Which angle is 180°? ________

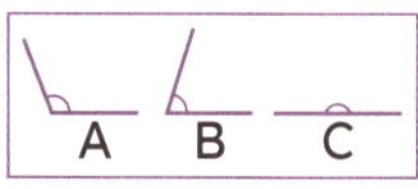

18. What is the perimeter? ________

11cm
(Not to scale.) 3cm

Monday

1. What is the place value of the 4 in 473?

1 ☐ 10 ☐ 100 ☐ 4 ☐

2. 37 × 5 = (30 × 5) + (7 × 5) = ______ + 35 = 185

3. What type of triangle is this?

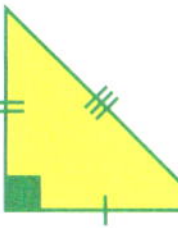

______

4. $1 + \frac{3}{10} =$ ______ (Write as a decimal.)

5. Draw a vertical line and label it as V, then draw a horizontal line and label it as H.

6. 21 – 9 = ______

7. Write ninety thousand and ninety as a numeral. ______

8. 2.1m =

21cm ☐ 210cm ☐
0.21cm ☐ 201cm ☐

9. Emily was counting by 100s. What did they count before 9090?

______

10. What is the perimeter of a square with sides 8cm long? ______

11. Which letter is not symmetrical:

I, F, or T? ______

12. Rotate a $\frac{1}{2}$ turn anticlockwise.

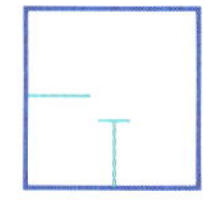

13. If today is 7 April, what will be the date in eight days time? ______

14. Bronte took 12 selfies each day for a week. (Write as a number sentence.)

______

15. Write the decimals in ascending order.

2.1 1.08 2.01 1.8 2.2 1.99

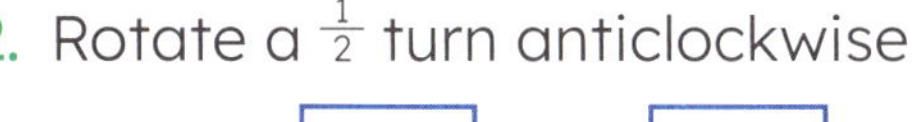

______ ______ ______ ______ ______ ______

Tuesday

Week 4

1. What is the value of the 9 in 397?

9 ☐ 90 ☐ 900 ☐ 10 ☐

2. Add a quarter of an hour to the time.

______

3. Chef paid $30 for onions priced at $6 per 10kg. How many kilograms of onions did Chef buy?

______

4. The product of 3 and 8 is ______.

5. 0.4, 0.6, 0.8, ______

6. What will be the length of A? ______m

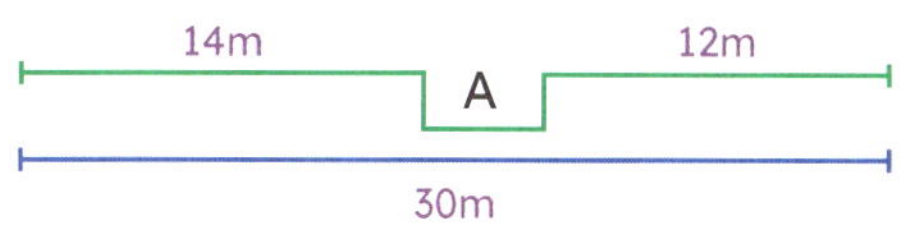

7. What type of triangle is this?

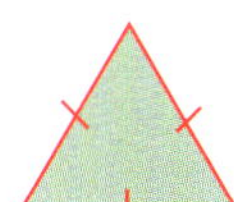

______

8. Complete the multiples of 11.

11, ______, ______, ______, ______, ______, ______, ______, ______, ______, ______, ______

9. 48 × 7 = (______ × 7) + (______ × 7)

= 280 + ______

= ______

10. Draw another line of symmetry.

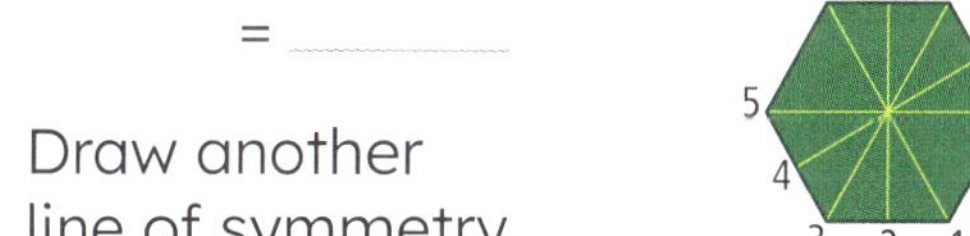

11. A mobile phone plan costs $30 each month. What is the total cost for a year? (Write as a number sentence.)

______

12. Round 9439 to the nearest 100.

______

13. 0.6 + 0.2 = ______

14. What is the number after 999?

______

15. Double 185. ______

Wednesday

Week 4

1. What is the place value of the 0 in 9.0?

   0 ☐ 0.1 ☐ 0.01 ☐ 10 ☐

2. Write the fractions in descending order.

   $\frac{2}{3}$ $\frac{1}{4}$ $\frac{2}{5}$ $\frac{3}{4}$

   ______ ______ ______ ______

3. 17 – 8 = ______

4. $2 + \frac{4}{100} = 2.04$

   $3 + \frac{5}{100} =$ ______

5. Add 15 minutes to this time. ______

6. 4500 + ______ = 20 000

7. 0.1 + 0.9 = ______

8. Rotate a $\frac{3}{4}$ turn clockwise.

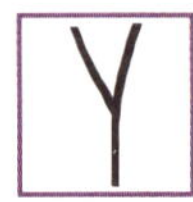
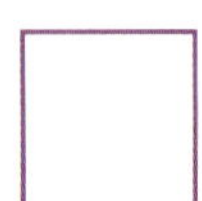

9. 2.2 < 2 true ☐ false ☐

10. 220 – 60 = ______

11. What type of triangle is this?

    ______

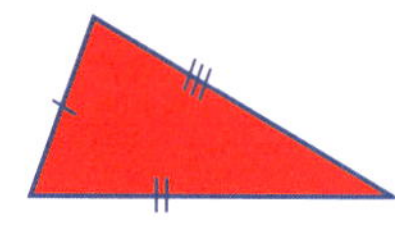

12. Write a number sentence for this problem: A jeweller has a box of 24 diamonds. If each row held eight diamonds, how many diamonds in each row?

    ______

13. If today is 17 January, what will be the date in one fortnight?

    ______

14. 1L = ______mL

15. Write the missing fractions.

Thursday

1. What is the value of the 7 in 2.7?

   7 ☐ 0.7 ☐ 0.07 ☐ 70 ☐

2. 15, 30, 45, ______, 75

3. What is the cost of 5kg of cashew nuts at $12.00 per kg?

   ______

4. 8 + 7 = ______

5. 1.1t =

   11kg ☐ 1100kg ☐
   101kg ☐ 1011kg ☐

6. 39 × 6 = (______ × 6) + (______ × 6)

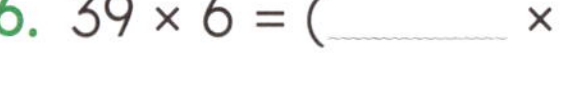

   = ______ + ______

   = ______

7. 1.1m =

   11mm ☐ 100mm ☐
   1100mm ☐ 1010mm ☐

8. What type of triangle is this?

   ______

9. What is the number after 799?

   ______

10. Which letter is symmetrical? ______

    Z S A

11. How many times will five go into 100?

    ______

12. $1 + \frac{7}{10}$ = ______ (Write as a decimal.)

13. Round 3031 to the nearest 100.

    ______

14. Which month is six months after July?

    ______

15. This is a

    ______.

Problem-solving

The bank made an error on Wenjing's recent deposit of $43.21 and mixed up the place value of the four numbers. The decimal point was in the correct place.

Let's find out the possible combinations.

Read the question again. **Think** about the information. Underline the important words.

Tick the strategy you will use to work out the answer:

- estimate and check ☐
- look for patterns ☐
- draw a diagram or picture ☐
- construct a table or graph ☐
- use materials ☐
- act it out ☐
- work backwards ☐
- something else. ☐

Solve it:

Reflect on the question and answer.

Check it. Circle another strategy on the list to work out the answer.

Show it:

Friday Review

Week 4

1. What is the value of the 3 in 2.3?

   1 ☐ 0.3 ☐
   0.03 ☐ 30 ☐

2. What is the length of A?

   ______ m

   13m A 12m
   30m
   (Not to scale.)

3. 0.1 + 0.9 = ______

4. Write the decimals in ascending order.

   3.2 3.03 3.1
   2.9 2.99

   ______ ______ ______
   ______ ______

5. 0.7, 0.8, 0.9, ______

6. Write a number sentence to solve the quantity. Alicia took eight selfies each day for a week.

   ______

7. 320 – 50 = ______

8. $2 + \frac{3}{100} = 2.03$

   $2 + \frac{8}{100} =$ ______

9. Round 3513 to the nearest thousand.

   ______

10. Write the number 100 before 7080.

    ______

11. 2.3 > 2

    true ☐ false ☐

12. How many kg of potatoes would you have if you paid $4.50 per 5kg and you spent $45.00?

    ______ kg

13. Write these fractions in descending order.

    $\frac{2}{5}$ $\frac{3}{4}$ $\frac{1}{2}$ $\frac{2}{3}$

    ______ ______ ______ ______

14. Add one-quarter of an hour to the time.

    ______

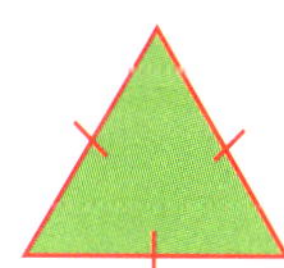

15. 2.1km =

    21m ☐
    210m ☐
    2100m ☐
    2010m ☐

16. Name this type of triangle.

    ______

17. Draw the other 2 lines of symmetry.

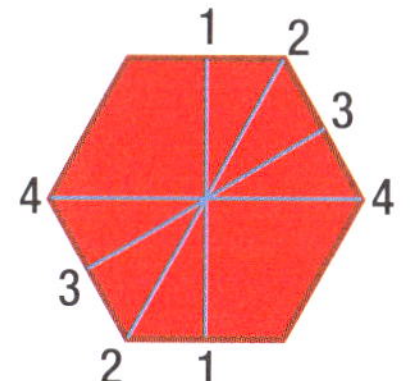

18. If the date is 31 March, what was the date a fortnight ago?

    ______

P

Week 5

Monday

1. 38 ÷ 6 = ______ r ______
2. 70 – 4 = ______
3. The decimal at: A = ______ B = ______

   C = ______ D = ______

   0 0.25 A B 1 C D 1.75
4. 93 × 3 = (______ × 3) + (3 × 3)

   = ______ + 9

   = ______
5. 4.2km = 42m ☐ 420m ☐ 4020m ☐ 4200m ☐
6. Round 3615 to the nearest thousand.

   ______
7. Measure the length of $\overline{AB}$ in cm.

   ______cm

   A B
8. What type of triangle is this?

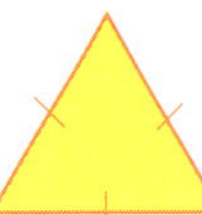

   ______
9. 6500 + ______ = 50 000
10. 3 + $\frac{8}{100}$ = ______ (Write as a decimal.)
11. What is the date one week before 7 October? ______
12. How many internal angles in an octagon? ______
13. Add 20 minutes to this time.

   7:10 ______
14. Write a number sentence to solve: There are three buckets, each holding 12 golf balls. How many golf balls altogether?

   ______
15. What is the perimeter?

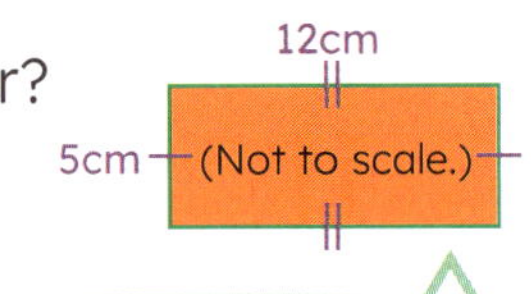

   ______

Tuesday

1. 25 ÷ 8 = ______ r ______
2. Add one-third of an hour to this time.

   11:25 ______
3. \$10.00 – \$8.70 = ______
4. Double 195. ______
5. Rotate a $\frac{1}{2}$ turn.

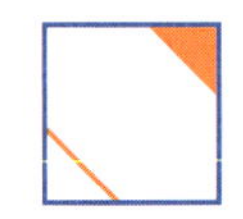

6. Alicia paid \$10.00 for 4kg of tasty potatoes. What is the cost per kg?

   ______
7. Read the pie graph and calculate the amount for green.

   ______
8. 0.9 + 0.1 = ______
9. Draw the third line of symmetry of this equilateral triangle.

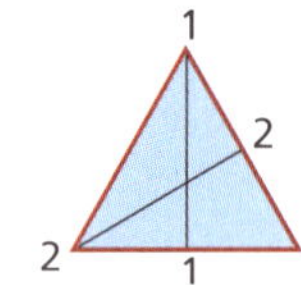

10. 57 917 – ______ = 50 000
11. What is the area of this 3 × 3 grid?

   ______ squares
12. 4.5 > 3.8 true ☐ false ☐
13. 0, 4, 8, ______, ______, ______, 24, ______
14. $\frac{1}{4} + \frac{2}{4}$ = ______
15. The year 2016 was a leap year. On this time line, complete the other leap years.

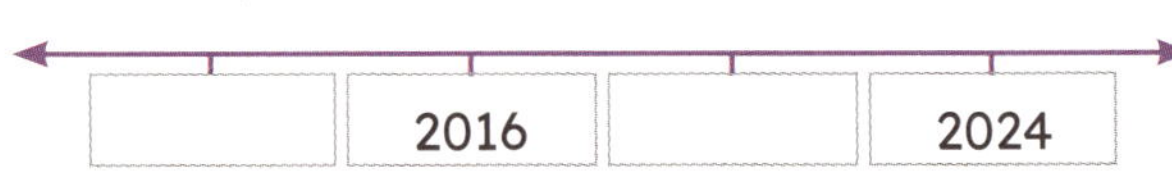

Wednesday

1. 41 ÷ 9 = ______ r ______

2. Write the numeral one thousand and ten. ______

3. Complete the multiples of 12.

   12, ______, ______, ______, ______, ______,

   ______, ______, ______, ______, ______, ______

4. 42 – 7 = ______

5. 7.2t = 7200kg ☐ 72kg ☐ 720kg ☐ 702kg ☐

6. What type of triangle is this?

   ______

7. 5.3 rounded to the nearest whole number is 5, so 7.4 rounded to the nearest whole number is

   ______.

8. 84 × 4 = (80 × 4) + (______ × 4)

   = ______ + 16

   = ______

9. Which mixed number is coloured?

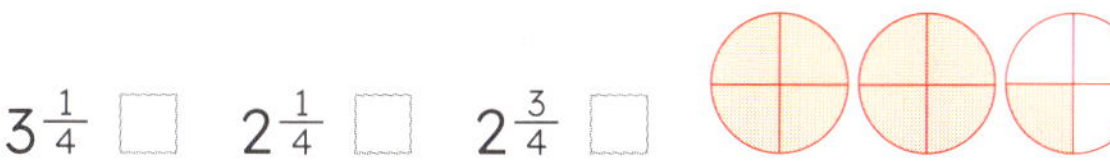

   $3\frac{1}{4}$ ☐ $2\frac{1}{4}$ ☐ $2\frac{3}{4}$ ☐

10. 80 000 + 70 + 5000 = ______

11. Which year was one century prior to 1984? ______

12. 0.6 + 0.4 = ______

13. Share $20.00 equally among eight students. ______

14. What is the area of this 4 by 3 grid?

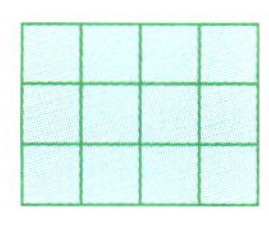

   ______ squares

15. What is the easier way of calculating the above area than counting all the squares?

   ______

Thursday

Week 5

1. 103 ÷ 10 = ______ r ______ = ______.______

2. Complete the decimals.

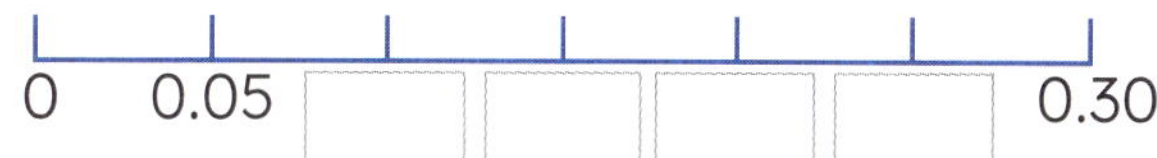

3. 999 + 7 = ______

4. This triangle is a

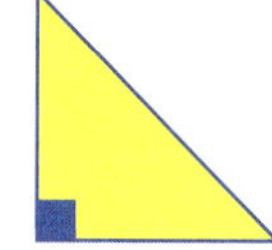

   ______

   ______.

5. 71 ÷ ______ = 7.1

6. 73 – 8 = ______

7. Round 8.7 to the nearest whole number. ______

8. In which season does January occur?

   ______

9. 1 – 0.8 = ______

10. 25, 50, 75, 100, ______, 150

11. 9.3 > 8.8 true ☐ false ☐

12. Colour the mixed number $3\frac{1}{3}$.

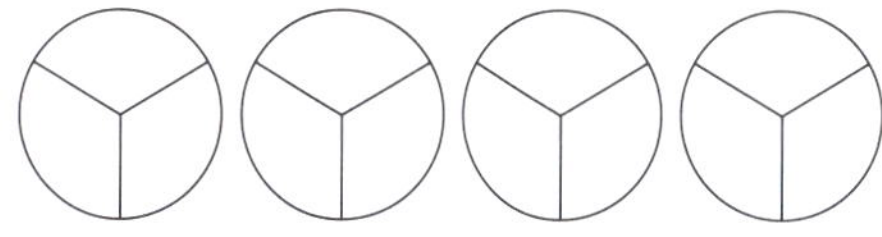

13. Using 7, 8, 4, and 8, write the greatest number possible.

   ______

14. 8 + 8 + 8 + 8 = ______

15. Read the pie graph and calculate how many people favour Daisy as a name.

   ______

Week 5

Problem-solving

Michele saw animal tracks every 16 metres on a 200-metre bush walk. They saw the first animal tracks 16 metres into the walk.

Let's find out how many whole tracks they saw and what metre of the walk Michele saw the last animal track. (Hint: Use digital tools to solve.)

Read the question again. **Think** about the information. Underline the important words.

Tick the strategy you will use to work out the answer:

- estimate and check ☐
- look for patterns ☐
- draw a diagram or picture ☐
- construct a table or graph ☐
- use materials ☐
- act it out ☐
- work backwards ☐
- something else. ☐

Solve it:

Reflect on the question and answer.

Check it. Circle another strategy on the list to work out the answer.

Show it:

Friday Review

1. ______, 9.5, 9.0, 8.5

2. What is the chance of picking a blue marble from a hat holding two blue and two red marbles?

______

3. $10.00 – $3.40

= ______

4. 74 × 4

= (70 × 4) +

(____ × ____)

= ______ + 16

= ______

5. $\frac{7}{10} - \frac{2}{10} =$ ______

6. 52 – 7 = ______

7. 2.9 < 3.2

true ☐ false ☐

8. 93 ÷ 9

= ______ r ______

9. Colour the mixed number $2\frac{2}{3}$.

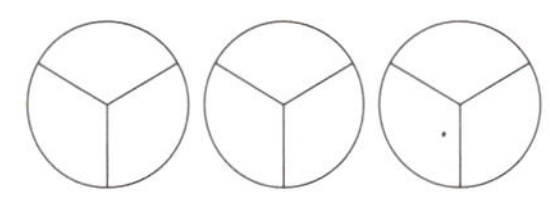

10. Using 6, 4, 1, and 3, write the largest number.

______

11. Round 7439 to the nearest thousand.

______

12. $3 + \frac{7}{100} =$ ______ (Write as a decimal.)

13. 36 ÷ ______ = 3.6

14. Add one-third of an hour.

______

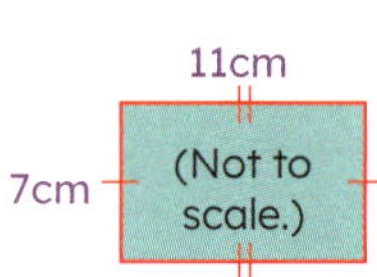

15. How many metres are in 3.7km?

307m ☐

3700m ☐

37m ☐

370m ☐

16. What is the perimeter of this rectangle?

______

17. Rotate a $\frac{3}{4}$ turn clockwise.

18. In Thursday's pie graph, how many more people prefer the name Toughie to Tom?

______

M

St

Monday

1. This is a net of a

________________.

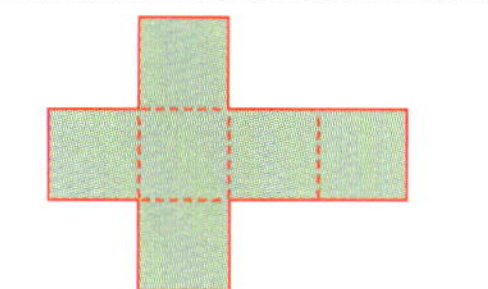

2. 200 + 250 + 300 + 250 = ________

3. Colour the improper fraction $\frac{8}{3}$.

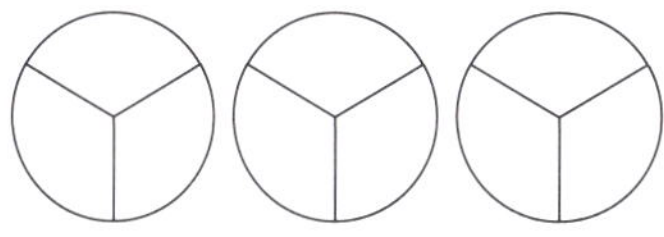

4. Draw a $2\frac{1}{2}$cm line and mark it as $\overline{AB}$.

5. \$10.00 – \$4.40 = ________

6. 0.5 + 0.7 = ________

7. Rotate a $\frac{1}{4}$ turn anticlockwise.

8. 2.4 = 2.0 + 0.4

   2.44 = 2 + 0.4 + 0.04

   3.88 = ________ + ________ + ________

9. 2300 – 900 = ________

10. odd × odd = ________

11. 3.01 < 3.10 true ☐ false ☐

12. Round 8650 to the nearest thousand.

____

13. Complete the multiples of 8.

8, ______, ______, 32, ______, ______, 56, ______, ______, 80

14. 4.7m = 470cm ☐ 47cm ☐ 407cm ☐ 4700cm ☐

15. Add a quarter of an hour to this time.

15:50 ________

Tuesday

Week 6

1. This is a net of a

____

________________.

2. What is the area of a grid which measures 2 squares by 5 squares?

______ squares

3. Write the number made by adding 1000 to eleven thousand and ten.

____

4. 7.7, 8.2, 8.7, 9.2, 9.7, ________

5. 6.37 = 6 + 0.3 + 0.07

   2.92 = ________ + ________ + ________

6. Round 7.4 to the nearest whole number.

______

7. What is the date one week after 24 June? ________________

8. Share \$50.00 equally among 4 people.

____

9. What is the place value of 8 in 12.8?

1 ☐ 10 ☐ 0.1 ☐ 0.01 ☐

10. $2 + \frac{4}{10}$ = ____.____

11. 9.5m = 95mm ☐ 905mm ☐ 9050mm ☐ 9500mm ☐

12. $\frac{8}{10} + \frac{1}{10}$ = ________

13. (9 × \$20) + (6 × \$10) = \$________

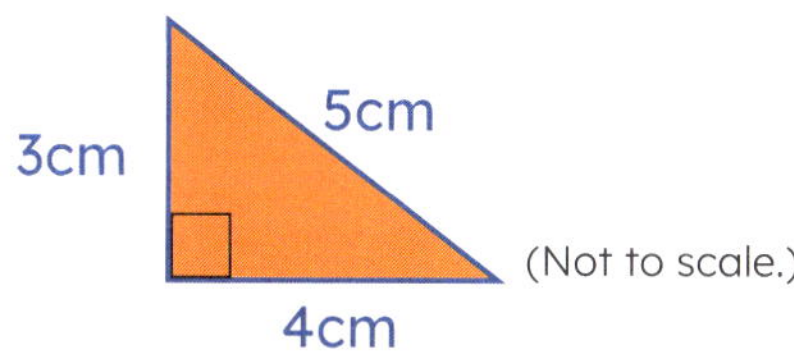

14. This right-angle triangle is also a ________________ triangle.

15. How many lines of symmetry does the triangle have? ______

Wednesday

Week 6

1. This is a net of a ______________ ______________.

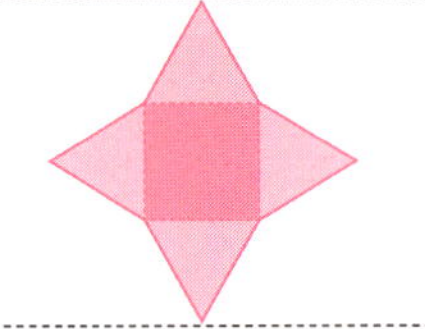

2. \$10.00 – \$9.70 = ______

3. 11 × 11 = ______

4. Rotate a $\frac{3}{4}$ turn clockwise.

5. What is the area of a 7 by 10 grid?

   ______ squares

6. Round 37.2 to the nearest whole number. ______

7. 0.9 + 0.7 = ______

8. 11 × 12 = ______

9. 7.5km = 750m ☐ 7500m ☐ 75m ☐ 705m ☐

10. What would you pay for 2kg of peas at \$9.50 per kg? ______

11. What is the number before 1890?

    ______

12. $5\overline{)70}$ = ______

13. 3 – 0.6 = ______, 3 – 0.06 = ______

14. (a) Which street is not in (B,1), (B,2), or (B,3)?

    ____

    (b) Write the coordinates for Erskine Street.

    ____

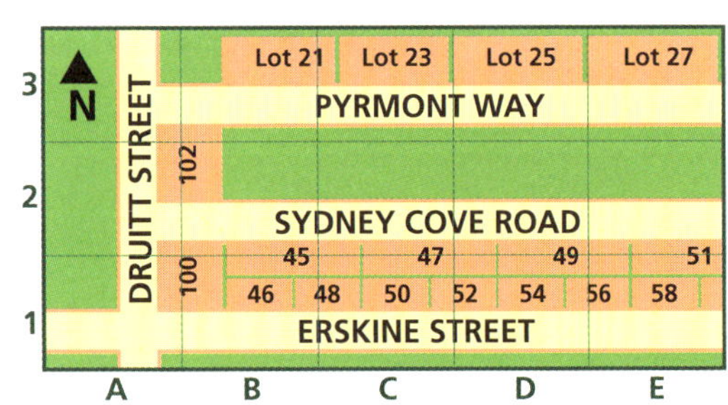

15. 24, 30, ______, 42, 48

Thursday

1. This is a net for a ______________.

2. Which equation is equal to 9 × 9?

   90 – 9 = 81 ☐

   80 + 9 = 89 ☐

   10 × 9 – 1 = 89 ☐

3. If you add five more tens to the number, what is the new amount?

   ______

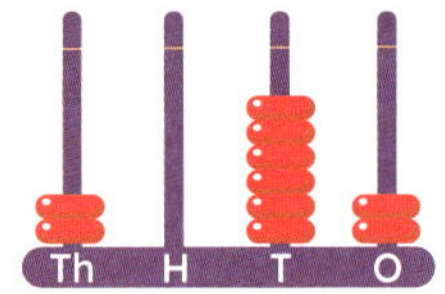

4. 3.19 = ______ + 0.______ + 0.______

5. 60, 90, 70, 100, 80, ______

6. Can an isosceles triangle be symmetrical? ______

7. 320 – 50 = ______

8. Add 15 minutes to this time.

   ______

9. 440 – 100 = ______

10. What is the perimeter?

    ______cm

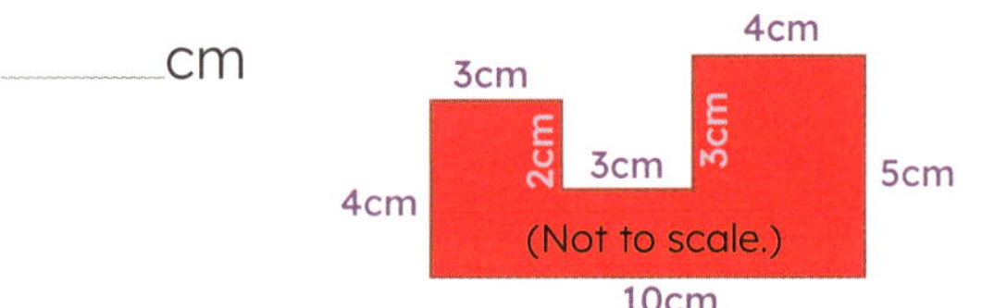

11. 82 × 5 = (______ × 5) + (______ × 5)

    = 400 + 10

    = ______

12. 700 + 600 + 1000 = ______

13. 17 + 28 = ______, 170 + 280 = ______

14. What is the probability of selecting a red marble from a bucket of eight blue and two red marbles?

    $\frac{2}{8}$ or $\frac{1}{4}$ ☐ $\frac{6}{8}$ or $\frac{3}{4}$ ☐ $\frac{2}{10}$ or $\frac{1}{5}$ ☐

15. 80 ☐ 50 = 130

Problem-solving

Kai plans to build a mouse box with five faces. The mouse will need to enter the box to get food.

Let's find out where the door could be created. (Hint: Look at the nets on page 115.)

Read the question again. **Think** about the information. Underline the important words.

Tick the strategy you will use to work out the answer:

- estimate and check ☐
- look for patterns ☐
- draw a diagram or picture ☐
- construct a table or graph ☐
- use materials ☐
- act it out ☐
- work backwards ☐
- something else. ☐

Solve it:

Reflect on the question and answer.

Check it. Circle another strategy on the list to work out the answer.

Show it:

Friday Review

Week 6

1. Round 9.4 to the nearest whole number.

_____

2. \$10.00 – \$7.80

= _____

3. ☐ ☐

0 ☐ ☐ 1.25

4. $5\overline{)90}$ = _____

5. 80 ☐ 40 = 120

6. 2, 10, 50, 250,

_____

7. Colour the improper fraction $\frac{7}{3}$.

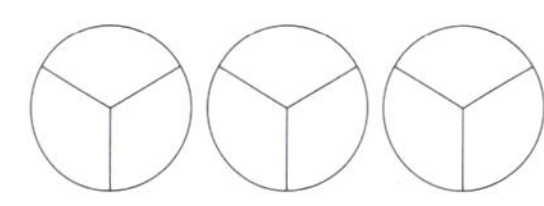

8. What is the area of a 5 by 6 grid?

_____ squares

9. Share \$100.00 equally among eight people.

_____ each

10. Write the multiples of 6.

72, 66, _____, _____,

_____, _____, _____

11. $3 + \frac{6}{10}$ = _____

(Write as a decimal.)

12. 2.01 < 2.10

true ☐ false ☐

13. Add one-quarter of an hour.

_____

7:55

14. Measure $\overline{AB}$.

_____ cm

A B

15. Look back at Wednesday's map. What is the lot number at (B,3)?

_____

16. After riding 3800m, Tim converted this distance to

_____ km.

17. If the date is 1 May, what was the date a fortnight ago?

_____

18. This is a net of a

_____

_____.

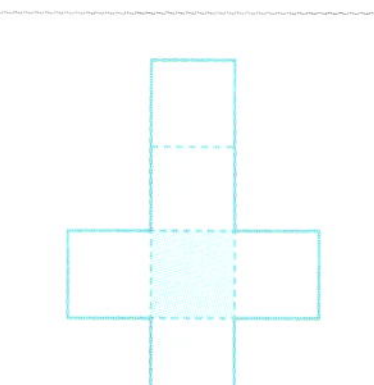

St

Week 7

Monday

1. $\frac{1}{2}$ of 20 = 10, so $\frac{1}{2}$ of 80 = ______.
2. 9035 – 100 = ______
3. How many months are there in two years? ______
4. Add half an hour to this time.

   5:20 ______
5. (5 × 18) + (5 × 14) = ______
6. 1000 – 10 = 990, so 10 000 – 10 = 9990, and 100 000 – 10 = ______.
7. The outcomes of flipping one coin are (H) or (T). The outcomes of flipping two coins could be:

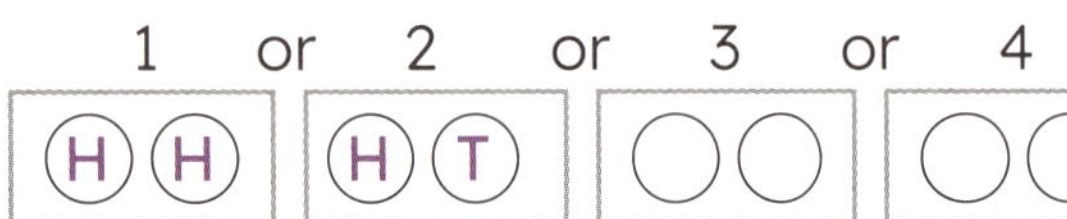

8. $5\overline{)100}$ = ______
9. This is a ______.

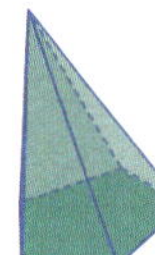

10. 0.8 = 8 hundredths ☐ 8 tenths ☐

    0.08 = 8 ______
11. Which triangle has three lines of symmetry?

    equilateral ☐ isosceles ☐ scalene ☐
12. The perimeter of this triangle is ______.

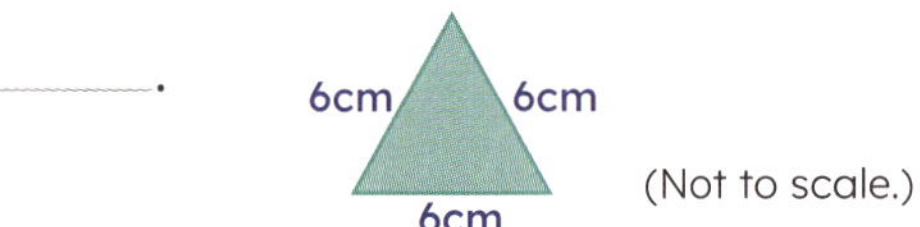

13. If a super-fast train travels at 400km per hour, how long would it take to travel 200km? ______
14. Write in descending order.

    1090 1019 1101 1001 1110

    ______ ______ ______ ______ ______
15. Round 6713 to the nearest thousand. ______

Tuesday

1. $\frac{1}{2}$ of 120 = ______
2. Write the next four multiples of 3.

   15, ______, ______, ______, ______
3. Write the missing factors of 24.

   1, 2, 3, ______, ______, 8, 12, 24
4. \$20.00 – \$12.50 = ______
5. 40 + 70 = ______
6. 870cm = 87m ☐ 8.7m ☐ 807m ☐ 870m ☐
7. 2 – 0.1 = ______
8. How many hours' difference is there between A and B?

   8 ☐ 6 ☐ 7 ☐ 11 ☐
9. $\frac{1}{4}$ = 0.25, $\frac{2}{4}$ = 0.50, $\frac{3}{4}$ = 0.75, $\frac{4}{4}$ = 1.0, $\frac{5}{4}$ = 1.25, $\frac{6}{4}$ = ______
10. 18 ☐ 10 = 1.8
11. *Fold paper in half.* *Cut shape.* *Unfold.*

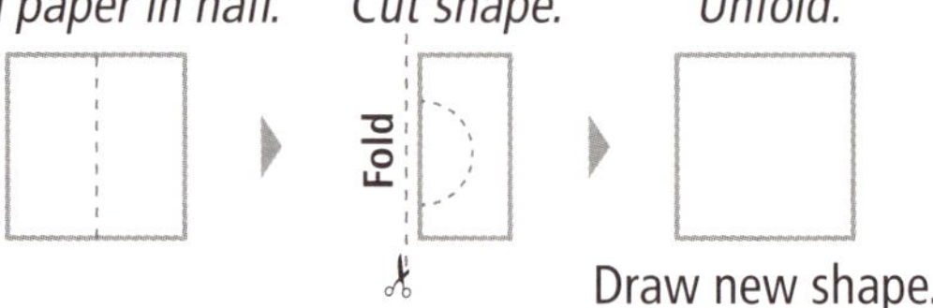

12. Which numbers are divisible by 4?

    75 ☐ 204 ☐ 305 ☐ 400 ☐
13. (3 × 9) + 3 = ______
14. 10 000 – 4600 = ______
15. What type of triangle is this? ______

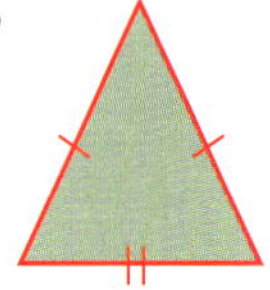

Wednesday

1. $\frac{1}{2}$ of 500 = ______

2. 0.6 + 0.4 = ______

3. If a fast train travels at 200km per hour and a journey from Meanjin / Brisbane to Canberra is a distance of 1200km, how long will the journey take?

______

4. 0, 9, ______, 27, 36, ______, ______, ______

5. Write the missing factors of 30.

1, 2, 3, ______, ______, 10, 15, 30

6. What is the probability of picking a king from a deck of 52 playing cards?

______

7. (a) 900 + 800 = ______

(b) 9000 + 800 = ______

8. If a bus left Tarndanya / Adelaide at 6 am and arrived in Port Fairy at 11 pm, how long was the journey?

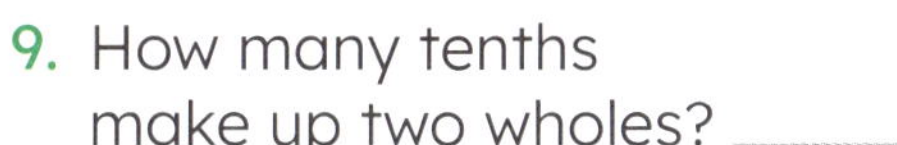

______

9. How many tenths make up two wholes? ______

10. 25 = 4 × 6 true ☐ false ☐

11. $\frac{1}{4}$ = 0.25

$\frac{2}{4}$ or $\frac{1}{2}$ = ______ (Write as a decimal.)

12. What is the cost of $2\frac{1}{2}$kg of tomatoes at $1.50 per kg?

______

13. $\frac{1}{4}$ < 0.75 true ☐ false ☐

14. If 9 × 9 = 81 (odd × odd = odd), does 11 × 11 = 121 or 122 (odd or even)?

______

15. This is a ______.

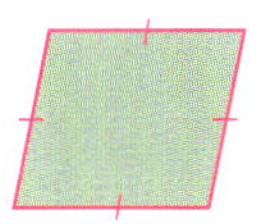

Thursday

Week 7

1. $\frac{1}{2}$ of 800 = ______

2. Add 30 minutes to this time.

______

3. What is the area of a 6 by 4 grid?

______ squares

4. The time difference is:

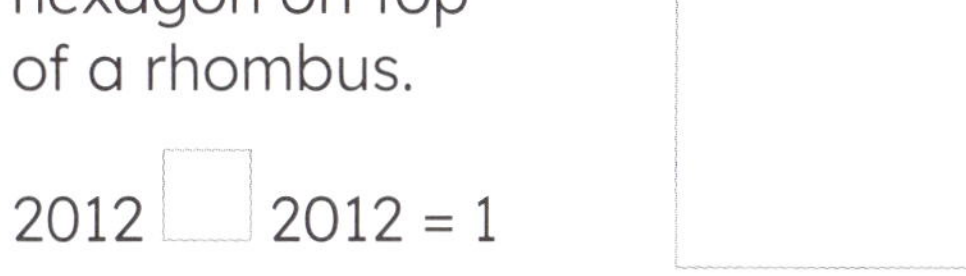

6 hours. ☐ 12 hours. ☐

18 hours. ☐ 36 hours. ☐

5. $6\overline{)84}$ = ______

6. Draw an irregular hexagon on top of a rhombus.

7. 2012 ☐ 2012 = 1

8. If odd × even = even, does 6 × 7 = 41 or 42? ______

9. What is the probability of an odd number?

______

10. 55mm = 550cm ☐ 5.5cm ☐ 505cm ☐ 50.5cm ☐

11. You buy a comic book for $8.60. What change do you receive from $20? ______

12. A rectangle has a rotational symmetry to the order of:

2 ☐ 3 ☐ 4 ☐

13. Round 16.9 to the nearest whole number. ______

14. How many fifths make up three wholes? ______

15. Write the next four multiples of 4.

28, ______, ______, ______, ______

Problem-solving

Week 7

Every month, Marv banks $50 into a savings account. At the end of every six months, the bank deposits half of any money added across that period as interest.

Let's find out how much interest Marv gets after saving for two years.

Read the question again. **Think** about the information. Underline the important words.

Tick the strategy you will use to work out the answer:

- estimate and check ☐
- look for patterns ☐
- draw a diagram or picture ☐
- construct a table or graph ☐
- use materials ☐
- act it out ☐
- work backwards ☐
- something else. ☐

Solve it:

Reflect on the question and answer.

Check it. Circle another strategy on the list to work out the answer.

Show it:

Friday Review

1. 1000 – 10 = 990
   10 000 – 10 = 9990
   100 000 – 10
   = ________

2. Write $\frac{1}{4}$ as a decimal.
   ____

3. 1717 – 17
   = ________

4. $\frac{1}{2}$ of 140 = ________

5. 0.8 + 0.2 = ________

6. 290 + 80 = ________

7. Round 2.7 to the nearest whole number.
   ____

8. 0.07 = ☐ tenths = ☐ hundredths

9. Write the next four multiples of 4.
   36, ________, ________, ________, ________

10. $5\overline{)140}$ = ________

11. 8 × 9 = ________

12. How many more Year 2 classes visited the zoo than Year 4 classes?
    ____

Number of visits to the zoo in a year

No. of classes: 0, 2, 4, 6, 8, 10

School year group: 0, 1, 2, 3, 4, 5, 6

13. 6800m =
    68km ☐
    6.8km ☐
    0.68km ☐
    680km ☐

14. Draw a vertical line next to an irregular pentagon.

15. Does a trapezium have rotational symmetry?

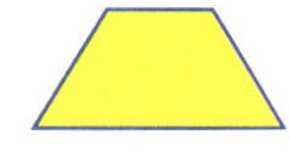

yes ☐ no ☐

16. How many hours are there from 2 pm to midnight?
    ____

17. The time difference is:

8 April 9 April

12 hours. ☐
6 hours. ☐
18 hours. ☐
20 hours. ☐

18. Using the spinner in Thursday, what is the probability of a multiple of 3?
    ____

P

Monday

1. Look at the spinner.
The possible outcomes are

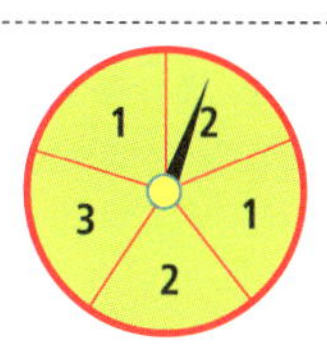

______________.

2. Add two-thirds of an hour to this time.

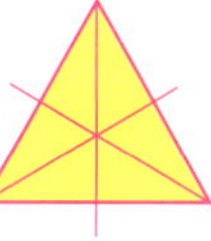

______

3. How many hours' difference from 3 pm to midnight?

______

4. 889 + 5 = ______

5. What type of triangle is this?

______ (Lines of symmetry.)

6. If odd × odd = odd,
does 13 × 13 = 169 or 170? ______

7. Draw the top view of this 3D object.

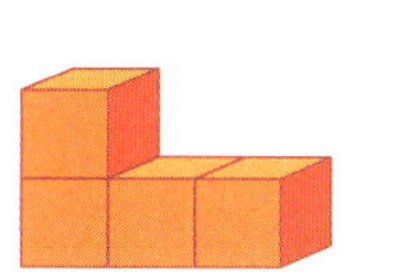

8. 300, 3000, 30 000, ______

9. Write nine-tenths as a decimal.

______

10. 50 × 8 = ______

11. Read the pie graph.
How many people liked juice?

______

12. Colour the prime numbers.

1 2 3 4 5 6 7 8 9
10 11 12 13

13. How many tens go into nine hundred?

______

14. 4.7km = 470m ☐ 4700m ☐ 47m ☐ 4.7m ☐

15. 230 – 90 = ______

Tuesday

1. What is the probability of landing on an odd number?

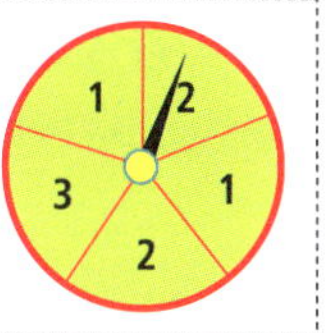

______

2. 8200 + ______ = 15 000

3. 491 ÷ ______ = 49.1

4. Lara built a 3D object using four equilateral triangles. Lara built a:

triangular cube. ☐ triangular prism. ☐
triangular pyramid. ☐ cone. ☐

5. $6\overline{)1020}$ = ______

6. Write the missing factors of 36.

1, 2, 3, 4, ______, 9, ______, 18, 36

7. There is a $4\frac{1}{2}$ hour difference between clocks A and B. For each clock, write if the time is am or pm.

A = 11 ☐m B = 3:30 ☐m

8. even – odd = ______

9. This triangle has one line of symmetry. What type of triangle is it?

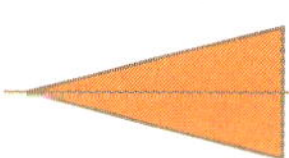

equilateral ☐ isosceles ☐ scalene ☐

10. 81 ☐ 9 = 9

11. How many odd-numbered houses are there in a street numbered from 1 to 14? ______

12. What is the length of this pencil?

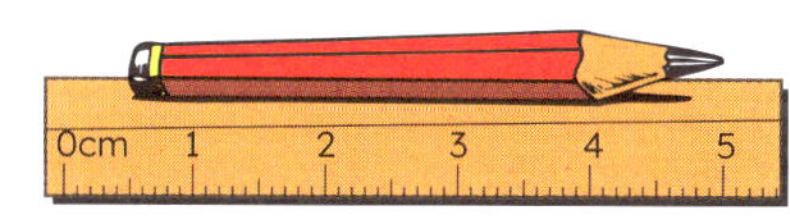

______mm

13. 1 – 0.9 = ______, 1 – 0.09 = ______

14. (a) 7 × 9 = ______ (b) 9 × 9 = ______

15. 2500, 500, 100, ______, 4

Week 8

Week 8

Wednesday

1. The outcomes of this spinner are:

______________________.

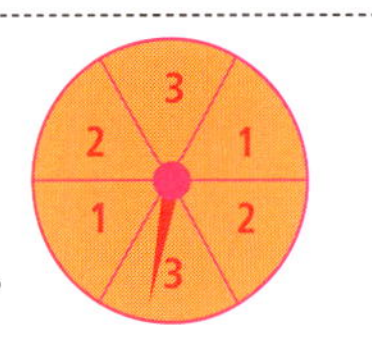

2. 17 ÷ 4 = ______ r ______

3. 5000, 1000, 200, 40, ______

4. 0.8 > 2 true ☐ false ☐

5. A is a ______________.

B is a ______________.

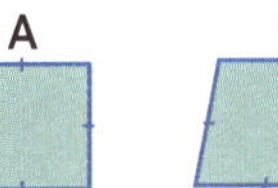
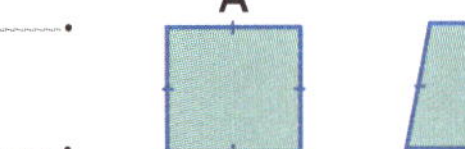

6. 40 000 + 60 000 = ____________

7. How many hours' difference is there from 11:30 am to 11 pm on the same day? ________

8. Draw the top view of this 3D object.

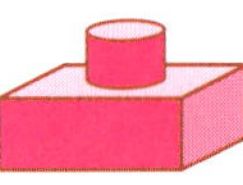

9. How many twenties go into two hundred? ______

10. Write the missing factor of 20.

1, 2, ______, 5, 10, 20

11. For your family car, how much does each tyre cost if the total cost for buying them is $600? **(Don't count the spare!)**

$________ each

12. 4.8kg = 4800g ☐ 480g ☐ 48g ☐ 4080g ☐

13. Complete the number line.

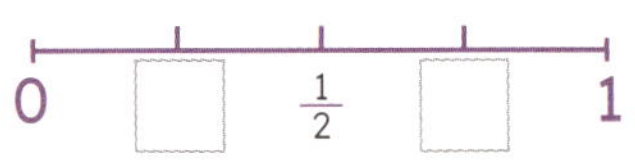

14. Is this an acute, obtuse, or right angle?

____

15. What is the area of a 7 by 9 grid?

______ squares

Thursday

1. The probability of a 2 is:

______.

2. Each unit on the number line represents:

$\frac{1}{4}$ ☐ $\frac{1}{9}$ ☐ $\frac{1}{8}$ ☐ $\frac{1}{2}$ ☐

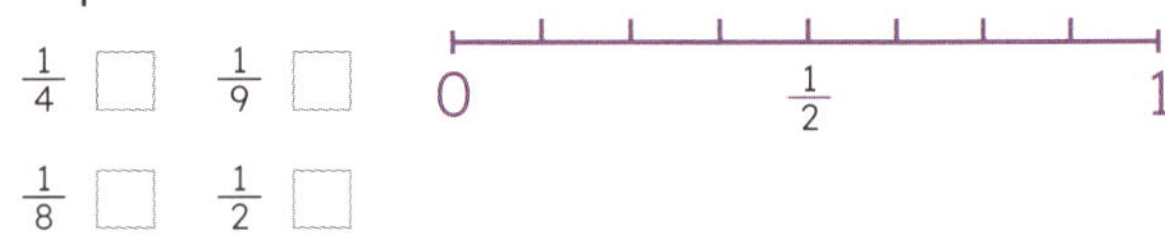

3. Which numbers are divisible by 5?

70 ☐ 32 ☐ 400 ☐ 195 ☐

4. Add 40 minutes to this time.

10:40 ____________

5. 2 – 0.4 = ________, 2 – 0.04 = ________

6. 4)2080 = ________

7. 900 000 – 700 000 = ____________

8. Write in descending order.

$\frac{2}{3}$ $\frac{1}{5}$ $\frac{3}{4}$ $\frac{1}{3}$

______ ______ ______ ______

9. Rotate a $\frac{3}{4}$ turn anticlockwise.

10. Write one hundred thousand as a numeral. ____________

11. If the perimeter of this triangle is 12cm, what length is A?

6cm ☐ 3cm ☐ 4cm ☐ 5cm ☐

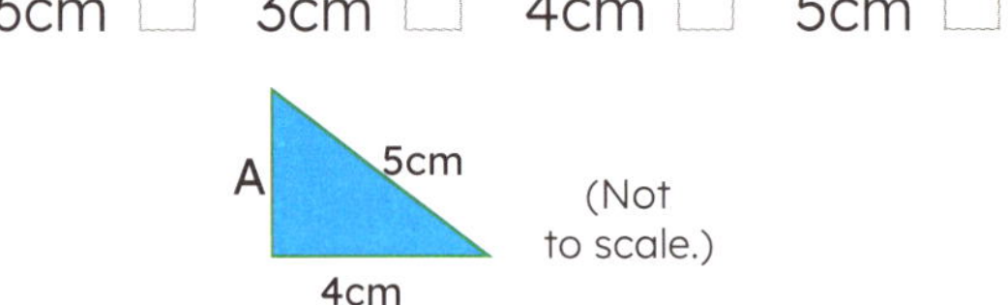

12. Write the next four multiples of 6.

36, ______, ______, ______, ______

13. 20 × 9 = 180, 19 × 9 = ________

14. 36 ÷ ______ = 2 × 3

15. 50 + 50 000 = ____________

Problem-solving

Gill was using a three-coloured spinner. During 100 spins, red came up 51 times, purple came up 24 times, and blue came up 25 times.

Let's find out what the possible outcomes were and what the spinner might have looked like.

Read the question again. **Think** about the information. Underline the important words.

Tick the strategy you will use to work out the answer:

- estimate and check ☐
- look for patterns ☐
- draw a diagram or picture ☐
- construct a table or graph ☐
- use materials ☐
- act it out ☐
- work backwards ☐
- something else. ☐

Solve it:

Reflect on the question and answer.

Check it. Circle another strategy on the list to work out the answer.

Show it:

Friday Review

Week 8

1. Each unit on the number line represents:

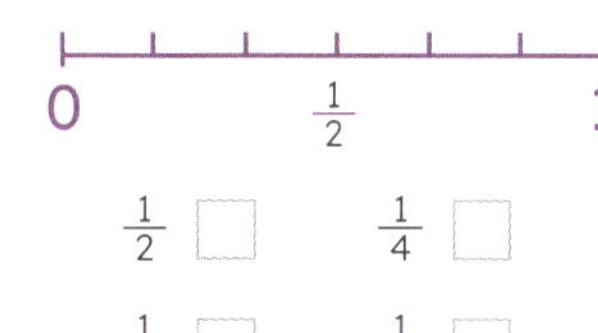

$\frac{1}{2}$ ☐ $\frac{1}{4}$ ☐

$\frac{1}{6}$ ☐ $\frac{1}{7}$ ☐

2. $6\overline{)906}$ = ...

3. \$20.00 – \$7.90 = ...

4. In Monday's pie graph, how many more people liked water than cola?

....

5. Write the next four multiples of 6.

24, ..., ..., ..., ...

6. 250, 500, ..., 1000, ...

7. $\frac{1}{5} > \frac{1}{4}$

true ☐ false ☐

8. 275 ÷ .... = 27.5

9. Write seven-tenths as a decimal.

....

10. 84 ☐ 12 = 7

11. 20 000 + 80 000 =

....

12. Add two-thirds of an hour.

....

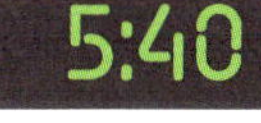

13. What type of triangle is this?

....

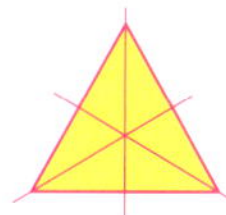

14. How many grams are in 2.9kg?

29g ☐ 290g ☐

2900g ☐ 209g ☐

15. Draw the top view.

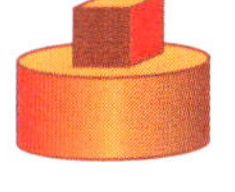

16. Is this an acute, obtuse, or right angle?

....

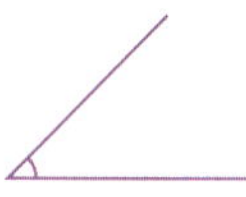

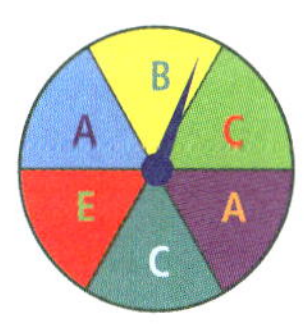

17. List the outcomes of the spinner.

....

18. The probability of landing on C is

....

P

Monday

Week 9

1. $\frac{9}{10} - \frac{4}{10}$ = ________

2. What is the value of the 9 in 9376?

____

3. Measure the length of $\overline{AB}$. ____cm

A ———————— B

4. What is the date one week before 6 April?

____

5. 1.2 ÷ 4 = 0.3, so 1.6 ÷ 4 = ________.

6. Round 6739 to the nearest 1000.

____

7. Aaron took seven selfies each day for a fortnight. What is the total number of pictures taken? (Write as a number sentence.)

____

8. 2900 + ________ = 5000

9. Add three beads to each place value and write the new value.

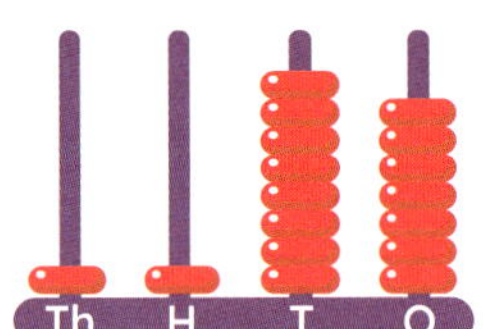

____

10. A reflex angle is an angle > 180° and < ________°.

11. Write the multiples of 25.

25, ____, ____, ____, ____, ____, ____, 200

12. If Moregatta is located 6km closer to the sign than Millaa Millaa, what distance needs to be written on the sign?

____

13. 10 – 0.1 = ________

14. This triangle has ______ line(s) of symmetry.

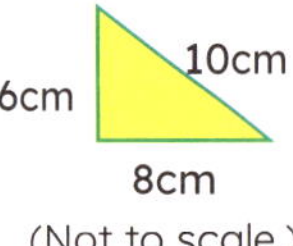

(Not to scale.)

15. 11 ☐ 4 = 44

Tuesday

1. $\frac{3}{10} + \frac{4}{10}$ = ________

2. Add one-third of an hour to this time.

9:50 ________

3. Match the quadrilateral names.

trapezium ____ square ____

kite ____ rhombus ____

4. 2 + 0.3 + 0.05 = ________

5. What is the value of the 5 in 5317?

____

6. What is the perimeter? ________

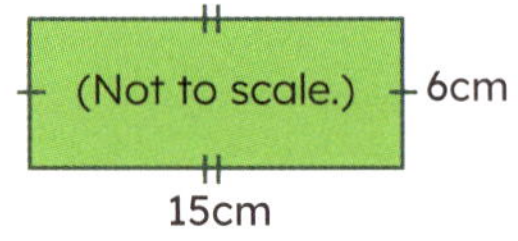

7. 200, 400, 800, 1600, ________

8. (a) 14 + 7 = ________

(b) 14 000 + 7000 = ________

9. 4 × 7 = ________

10. How many lines of symmetry does this shape have?

____

11. Write the missing factors of 50.

1, 2, ______, ______, 25, 50

12. odd + even = ________

13. If the sun is on the western horizon, is it am or pm? ________

14. 48mm = 480cm ☐ 4.8cm ☐ 4800cm ☐ 48cm ☐

15. How many 20g packets of potato chips would one 200g box contain?

____

Wednesday

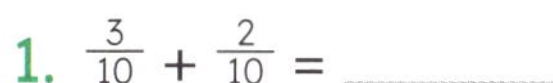

1. $\frac{3}{10} + \frac{2}{10} =$ ______

2. Write fifty thousand and five as a numeral.

______

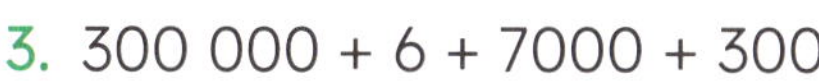

3. 300 000 + 6 + 7000 + 300

= ______

4. 4.1 – 0.3 = ______

5. Rotate a $\frac{1}{2}$ turn clockwise.

6. 4 × 8 = ______

7. Write the next four multiples of 4.

36, ______, ______, ______, ______

8. How many 20c coins make up $5?

______

9. What is the date one week from 26 June?

______

10. *Fold paper in half.* ▸ *Cut shape.* ▸ *Unfold.*

Draw new shape.

11. $\frac{1}{4} = 0.25$, $\frac{2}{4} = 0.5$, $\frac{3}{4} =$ ______

12. How many kilograms of sugar would you receive if you spent $25.00 and sugar costs $5.00 per 15kg?

______

13. 0.9 ÷ 3 = 0.______

14. Work out the GST and the total price for the piano repairers' invoice.

Stanislaus Bros.
Piano repairers

Repaired black and white keys	$300
+ GST at 10%	**A**
Total	**B**

15. Share $120.00 equally among 5 people. ______

Thursday

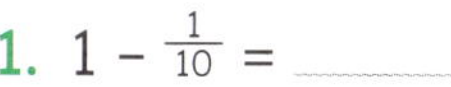

1. $1 - \frac{1}{10} =$ ______

2. 41.8 ☐ 10 = 418

3. Draw three more lines of symmetry.

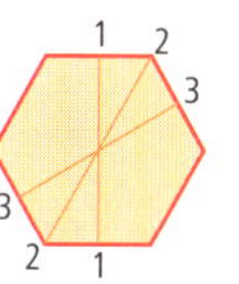

4. Measure the length of $\overline{ABC}$. ______cm

5. The perimeter of a regular pentagon with 9cm sides is

______.

6. A trainee architect was given this tape measure. Write the measurements below.

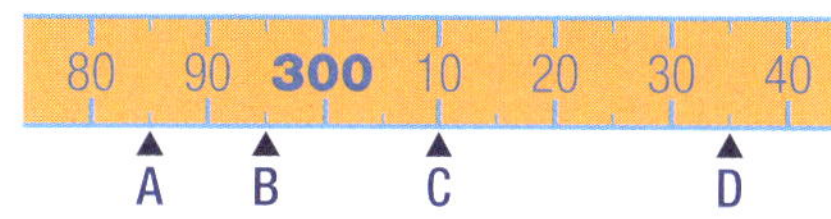

A = ______mm B = ______mm

C = ______mm D = ______mm

7. 6.2t = 620kg ☐ 62kg ☐ 6200kg ☐ 602kg ☐

8. 0.7 + 0.4 = ______

9. This is a net of a

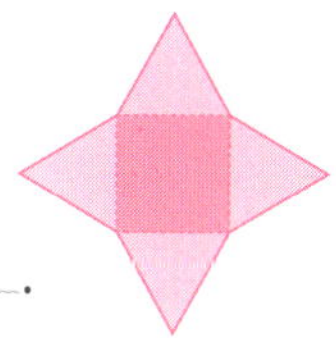

______.

10. $\frac{1}{3}$ of 150 = ______

11. Round 6.7 to the nearest whole number. ______

12. 350, 700, 1400, 2800, ______

13. What is the chance of the pointer landing on:

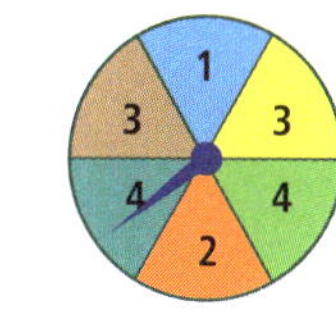

(a) a 3? ______

(b) a 1? ______

(c) an odd number? ______

14. 210 – 70 = ______

15. Add a half-hour to this time.

8:40 ______

Week 9

Problem-solving

Week 9

Friday is pizza night for Nanna, Dad, Charly, Jay, and Dee. The pizzas are sliced into quarters. Everyone wants $\frac{2}{8}$ each, but Dad wants $\frac{4}{8}$.

Let's find out how many whole pizzas are needed and what is left over. Write the left overs as a decimal.

Read the question again. **Think** about the information. Underline the important words.

Tick the strategy you will use to work out the answer:

- estimate and check ☐
- look for patterns ☐
- draw a diagram or picture ☐
- construct a table or graph ☐
- use materials ☐
- act it out ☐
- work backwards ☐
- something else. ☐

Solve it:

Reflect on the question and answer.

Check it. Circle another strategy on the list to work out the answer.

Show it:

Friday Review

1. $\frac{3}{4} + \frac{1}{4} =$ ______
2. $0.6 \div 2 =$ ______
3. $8.45 =$ ______ + ______ + ______
4. $\frac{1}{3}$ of $90 =$ ______
5. Add two beads to each place value and write the new amount.

______

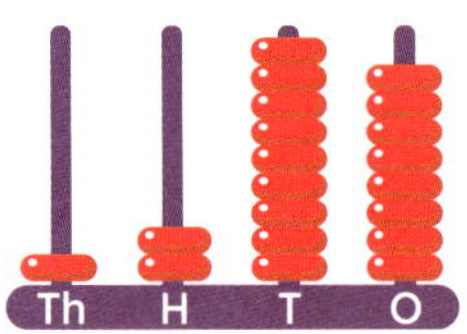

6. $20.00 – $11.10 = ______
7. Share $120.00 among 6 people.

______

8. 100 000 – 1000 =

______

9. Chef paid $30 for sugar. The sugar cost 20c per kg. How many kilograms of sugar did Chef buy?

______

10. odd × odd = ______
11. Round 9.7 to the nearest whole number.

______

12. What is the area of a 7 by 6 grid?

______ squares

13. Add one-quarter of an hour.

______

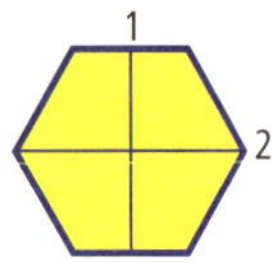

14. Draw the four remaining lines of symmetry.

1
2

15. If Cockatoo Valley is 7km further than Barossa Valley, what distance needs to be written on the sign?

______

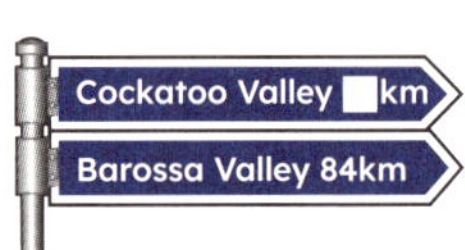

16. On this tape measure, what is the value of:

70 80 90 **700** 10
B A

A? ______ mm

B? ______ mm

17. Rotate a $\frac{3}{4}$ turn clockwise.

18. Look back at Thursday's spinner. List the outcomes.

______

N A M Sp St P

Week 10

Monday

1. If ↑ is north, then ↗ is ______________.

2. What would be the time in 45 minutes? ______

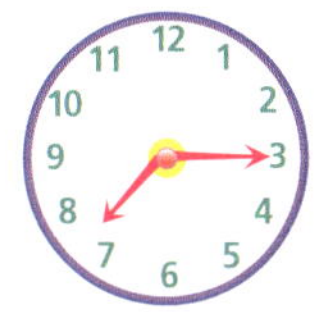

3. Your shopping trolley has six grocery items ranging in price from $6 to $10. The estimated total is:

   > $100 ☐ < $60 ☐ $760 ☐

4. The magic square sums to 15 (using 1 to 9) in all directions. Complete the square.

6		2
1	5	9
	3	

5. Write the date 4 May 2025 in numerals. ______

6. 100 – 70 = ______, 1000 – 70 = ______

7. 12 = 4 × 4 ☐ 3 × 4 ☐ 2 × 7 ☐ 5 × 3 ☐

8. Read the pie graph. How many students were late to school? ______

Survey of school punctuality

9. Write the next four multiples of 8.

   24, ______, ______, ______, ______

10. 4 × 8 = 16 + ______

11. Which two numbers are not composite?

    2 4 8 9 11 15

    ______ and ______

12. Measure the length of $\overline{CD}$. ______ cm

    C ______________________ D

13. The place value of 8 in 2.8 is ______.

14. The equilateral triangle has an order of rotational symmetry of: 2 ☐ 3 ☐

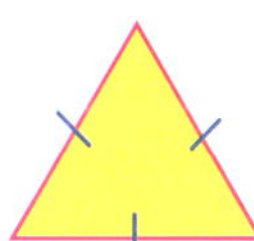

15. Write ninety thousand, nine hundred as a numeral.

    ______

Tuesday

1. If ↑ is north, then ↓ is ______________.

2. What would the time be in three-quarters of an hour? ______

   2:30

3. 9 + 6 = ______, 9 + 0.6 = ______

4. 210 – 30 = ______

5. 7 × 6 = 21 + ______

6. In one year, how many months have 30 days? ______

7. Colour the improper fraction of $\frac{8}{3}$.

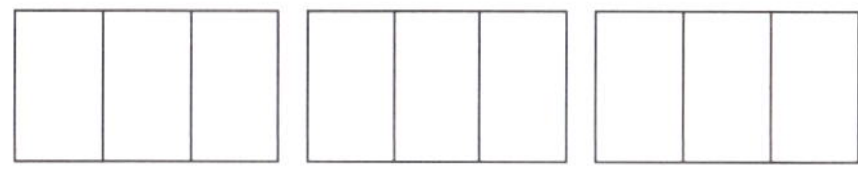

8. What is the value of the 7 in 7352? ______

9. 1.5 ÷ 5 = ______

10. Double 15. ______

    Halve 15. ______

11. 110, 220, 330, ______, 550

12. The outcomes of the spinner are ______________.

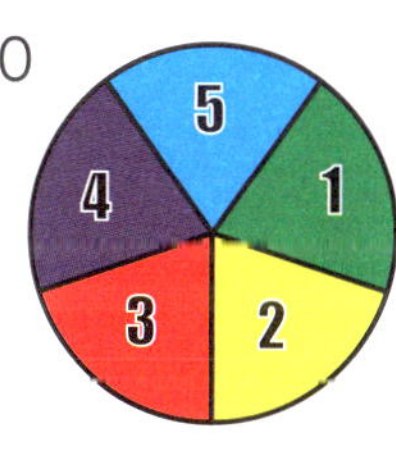

13. What is your weekly rate, if you received a total of:

    (a) $20 pocket money over four weeks? ______

    (b) $30 pocket money over four weeks? ______

14. $\frac{1}{3}$ of 63 = ______

15. In a 400g box there are 20 packets of potato chips. What is one packet's mass? ______

Wednesday

1. If ↑ is north, then → is ________.

2. Colour to show 1700mL.

3. $3\overline{)624}$ = ________

4. 1700 – 900 = ________

5. A super-fast train travels at 400km/h. A journey from Boorloo / Perth to Jerramungup is a distance of 400km. How long will the trip take if a super-fast train is built?

____

6. In a year, how many months have 31 days? ________

7. Draw the missing two lines of symmetry.

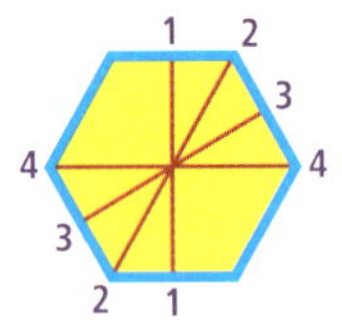

8. (a) 1000 – 5 = ________

   (b) 10 000 – 5 = ________

9. $\frac{1}{4}$ of 80 = ________

10. 2.1 ÷ 3 = 0.________

11. $20 – $16.70 = ________

12. 6 + 2 tenths + 4 hundredths = ________

13. Does a scalene triangle have rotational symmetry?

____

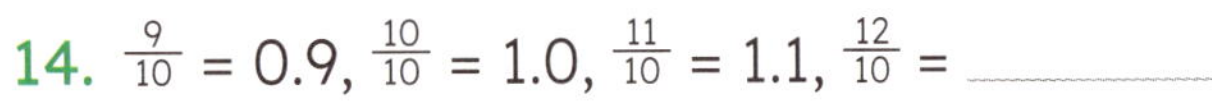

14. $\frac{9}{10}$ = 0.9, $\frac{10}{10}$ = 1.0, $\frac{11}{10}$ = 1.1, $\frac{12}{10}$ = ________

15. 6 × 6 = 18 + ________

Thursday

1. If ↑ is north, then ↙ is ________.

2. Measure the length of $\overline{XYZ}$. ________cm

3. In this set, which number is not a composite? ________

   15 18 19 20 21

4. Write the date 29 May 2017 in numerals. ________

5. Double 1950. ________

6. A reflex angle is between 180° and ________°.

7. What is the probability of choosing a:

   (a) green marble? ________

   (b) purple marble? ________

8. 5 + 7 tenths + 3 hundredths = ____.____

9. Write the next four multiples of 6.

   18, ________, ________, ________, ________

10. 50 × 28 × 2 = ________

11. The angles at A, B, C, and D are equal.

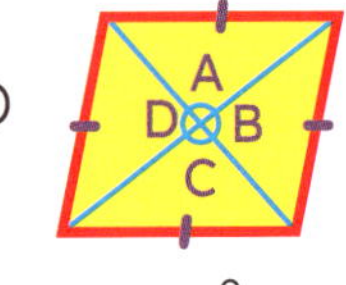

________° + ________° + ________° + ________°

12. How many leap years occurred from 1992 (a leap year) to 2015? ________

13. 6 + 9 = ________, 6 + 0.9 = ________

14. What is the length of A? ________cm

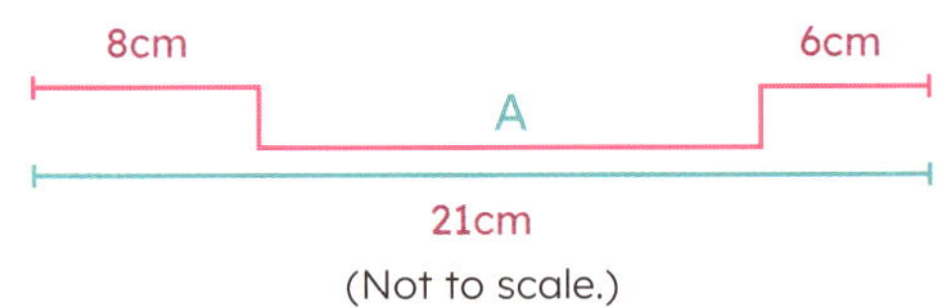

(Not to scale.)

15. $2 + \frac{3}{100}$ = ____.____

Problem-solving

Look at the map below in this column.

On Inka's trip around Australia, they started in Warrane / Sydney. They travelled north until they arrived at a capital city, then drove anticlockwise around the coast until they arrived in another capital city. Inka then flew to the furthest capital city away.

Let's find out the grid coordinates for the capital city where Inka ended the trip.

Read the question again. **Think** about the information. Underline the important words.

Tick the strategy you will use to work out the answer:

- estimate and check ☐
- look for patterns ☐
- draw a diagram or picture ☐
- construct a table or graph ☐
- use materials ☐
- act it out ☐
- work backwards ☐
- something else. ☐

Solve it:

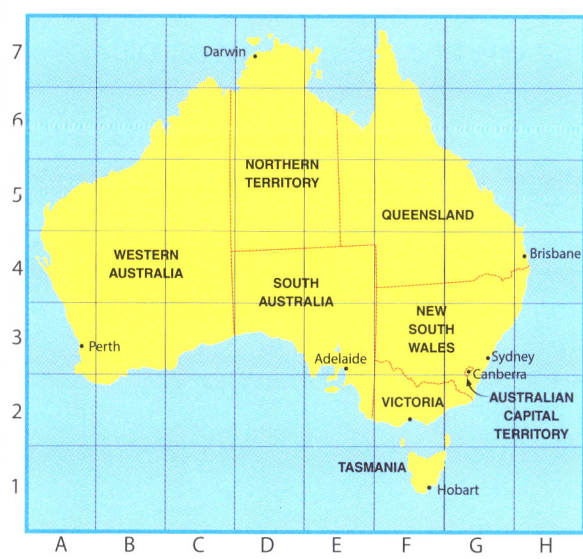

Reflect on the question and answer.

Check it. Circle another strategy on the list to work out the answer.

Show it:

Friday Review

1. $6 \times 9 = 27 +$ ______
2. $1000 - 6 =$ ______
3. 40, 80, 120, ______, 200
4. ______ $\div 4 = 100 \div 2 = 50$
5. $\frac{1}{4}$ of 80 = ______
6. The magic square sums to 15 (using 1 to 9) in all directions. Complete the square.

		2
	5	
8		4

7. 4 + 9 tenths + 3 hundredths = ____.____
8. $57 \times 50 \times 2$ = ______
9. Double 17. ______
10. $3 + \frac{3}{100} =$ ______ or ____.____
11. What is the value of the 9 in 8.9? ______
12. Write seven thousand, four hundred and eight as a numeral. ______
13. This is a ______.

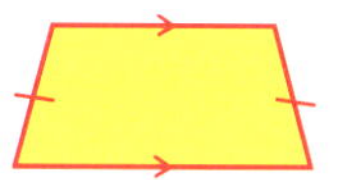

14. If ↑ is north, then ↖ is ______.
15. A rectangle has an order of rotational symmetry of ______.

16. If 2004 was a leap year, was 1994 or 1996 a leap year? ______
17. The probability of choosing a yellow is ______ in ______.

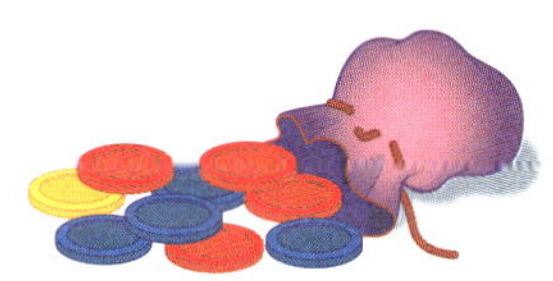

18. Read the pie graph. How many people watch TV? ______

TV viewing habits

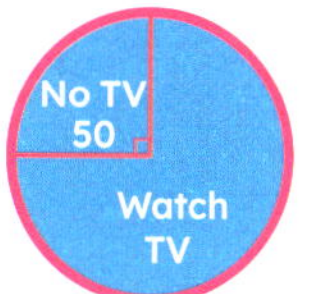

Week 10

P

Week 11

Monday

1. 50 ☐ 10 = 5

2. Measure the length of $\overline{ABC}$. ______ cm

3. 1m = 1000mm, so 4m = ______ mm.

4. What is the perimeter of a 30m by 15m rectangular building?

______

5. $5\overline{)345}$ = ______

6. Write ten thousand as a numeral.

______

7. 1 – 0.3 = ______

8. 

= $______

9. $\frac{1}{4}$ of 40 = ______

10. What is this 3D object?

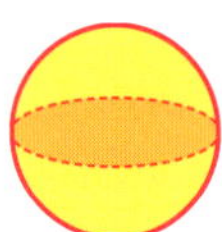

______

11. What is the value of the 1 in 12 400?

______

12. 7 ÷ 2 = 3r1

$\frac{7}{2} = 3\frac{1}{2}$ or 3.______

13. What was the time one quarter of an hour before?

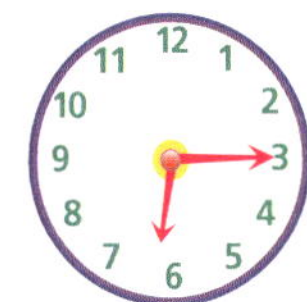

______

14. The magic square sums to 15 (using 1 to 9) in all directions. Complete the square.

		8
9	5	
2	7	

15. 0, $\frac{1}{2}$, 1, $1\frac{1}{2}$, ______

Tuesday

1. 12 ☐ 4 = 48

2. Round 8825 to the nearest thousand.

______

3. Circle the square number.

2 4 6 8 10

4. 61.4 × ______ = 614

5. Write A at 0.2.
Write B at 0.35.
Write C at 0.4.

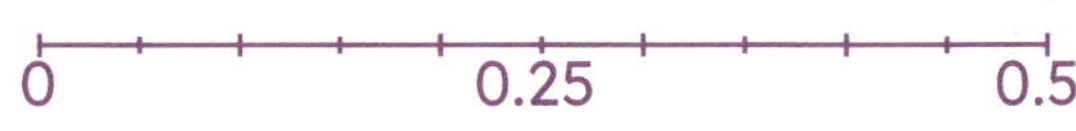

6. Write the date 7 April 2027 in numerals.

______

7. ______, 9.5, 9.0, 8.5, 8.0

8. 1 hectare = ______ m^2

9. If you own a bicycle shop and have stock of 117 bicycles, how many wheels are there (no spares)?

______

10. This abacus shows 12 132.
Add one bead to each place value and rewrite the number.

______

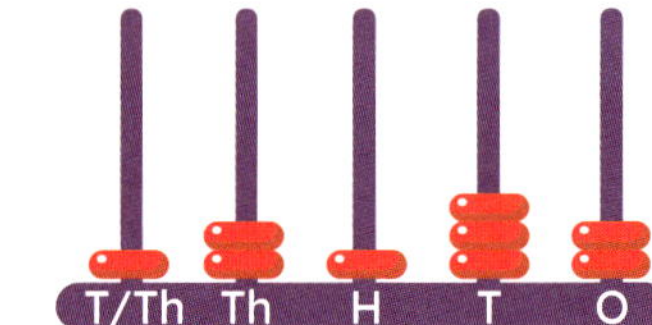

11. Using the digits 6, 0, 4, and 2, what is the lowest whole number you can arrange?

______

12. 130 – 70 = ______

13. Share $20.00 equally among 8 people.

______

14. $5\overline{)21}$ = 4r1 = $4\frac{1}{5}$ $6\overline{)25}$ = 4r1 = 4______

15. Is the remainder 1 the same for these divisions? ______

Wednesday

1. $5\overline{)\square}$ = 40

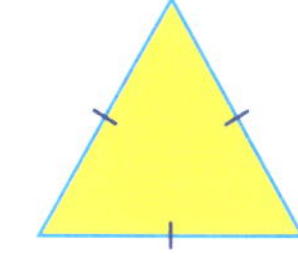

2. On this triangle, draw all the lines of symmetry.

3. Round 13 585 to the nearest thousand. ________

4. Write ten thousand and eleven as a numeral. ________

5. What is the direction if you:

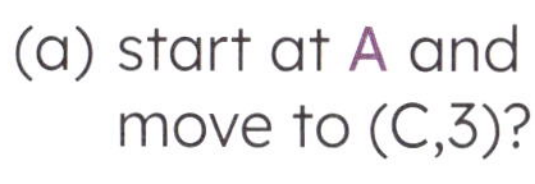

(a) start at A and move to (C,3)? ________

(b) move from (C,3) to (A,5)? ________

(c) move from (A,5) to (H,5)? ________

6. (a) 10 × 9 = ________ (c) 17 × 9 = ________

(b) 7 × 9 = ________

7. $\frac{15}{2}$ = ________ (mixed number)

= ________ (decimal)

8. Measure the length of $\overline{XYZ}$.

________ cm

9. 2.0 ÷ 4 = ________

10. What is the perimeter of an office building that is shaped as a regular pentagon with 10m sides?

____

11. 264 905 – ________ = 200 000

12. Calculate the amounts for the Marty's hardware docket.

Door hinges 2 @ $1.45	A
'Hard hit' hammer	$85
Total	B

13. Using 3, 7, 4, and 0, what is the greatest odd number you can arrange? ________

14. Which set has two prime numbers?

1, 5 ☐ 2, 9 ☐ 2, 7 ☐ 2, 15 ☐

15. 9.7, 9.8, 9.9, ________

Thursday

Week 11

1. 789 ÷ ________ = 78.9

2. What was the time 15 minutes before?

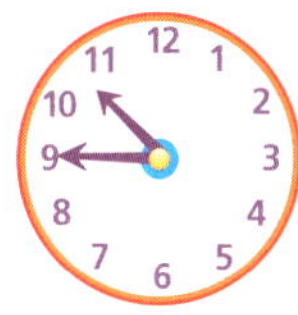

____

3. 48 × 7 = (________ × 7) + (________ × 7)

= ________ + ________

= ________

4. A square has a rotational symmetry to the order of: 2 ☐ 3 ☐ 4 ☐

5. Round 15 500 to the nearest ten thousand. ________

6. Tradie Tam measured lengths of pipe from zero to the four points marked. Record the measurements in mm.

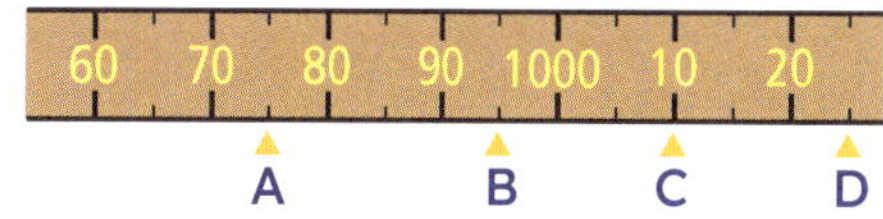

A = ________ mm B = ________ mm

C = ________ mm D = ________ mm

7. 3 + 0.4 + 0.01 = ________

8. If ↑ is north, then ↖ is

________.

9. $\frac{13}{5}$ = ________ (mixed number)

= ________ (decimal)

10. In this 500g box there are 20 packets of chips. What is one packet's mass?

________ g

11. 4 + 4 + 4 = 3 × ________ = ________

12. (a) 60 × 10 = ________

(b) 600 × 10 = ________

13. Double 0.9. ________

14.  = $________

15. 789 ÷ ________ = 78.9

Week 11

Problem-solving

It was Jen's grandad's birthday, and he celebrated by listening to 20 new songs. Each song played for 3 minutes.

Let's find out what fraction of his day was spent listening to the new songs.

Read the question again. **Think** about the information. Underline the important words.

Tick the strategy you will use to work out the answer:

- estimate and check ☐
- look for patterns ☐
- draw a diagram or picture ☐
- construct a table or graph ☐
- use materials ☐
- act it out ☐
- work backwards ☐
- something else. ☐

Solve it:

Reflect on the question and answer.

Check it. Circle another strategy on the list to work out the answer.

Show it:

Friday Review

1. $17 \div 2 = \frac{17}{2}$

   = ______ (mixed number)

   = ______ (decimal)

2. Halve 1. ______ . ______

3. 214 ÷ ______ = 21.4

4. 3000, 6000, 10 000, 15 000, ______

5. 1.6 – 0.7 = ______

6. 5)‾☐‾ = 120

7. Calculate the amounts for this hardware store docket.

Sandpaper 4 @ $1.10	A
Paint roller	$85
Total	B

8. Using the digits 9, 6, 0, and 7, create the lowest value possible.

   ______

9. Share $15.00 among 10 people.

   ______ each

10. Round 14 600 to the nearest ten thousand.

    ______

11. 2.0 ÷ 4 = ______

12. 

    = $______

13. What was the time one-quarter of an hour before?

    ______

    5:15

14. Which triangle has a rotational symmetry to the order of three?

    ______

    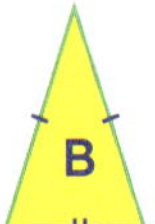

    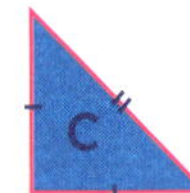

15. 3.5m = ______ mm

16. If ↑ is north, then ↙ is

    ______.

17. Name this object.

    ______

    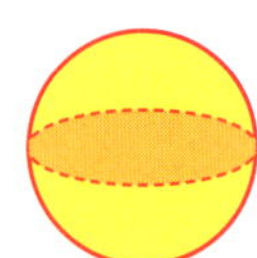

18. A 500g box holds 25 packets. What is each packet's mass?

    ______

Sp

P

Monday

1. How many hours' difference is there between A to B?

A	Date: 12 April
	Time: 2 hours past midnight

B	Date: 13 April
	Time: 1 hour past midnight

______

2. What 2D shapes can be found in a triangular prism?

______

3. $\frac{1}{2}$ of 52 = ______

4. Round $7\frac{3}{7}$ to the nearest whole. ______

5. Complete the number line.

6. 40 × 9 = 360, 39 × 9 = 351, 38 × 9 = 342,

   37 × 9 = ______

7. Colour a pair of parallel lines on this rectangle.

8. (7 × 2) – 9 = ______

9. WA is ______ hours behind NSW, Vic., Tas., and Qld. (Non-daylight saving time.)

10. Using the digits 2, 5, 3, and 0, what is the highest value you can create?

______

11. Draw a reflection (mirror image) of the letter shape.

12. 25 + 27 = ______

13. Hannah spent $3.40 on an ice cream. The change from $10 is

______.

14. $5\overline{)800}$ = $10\overline{)\quad}$ = 160

15. 5, 50, 500, ______

Tuesday

1. How many hours' difference is there between A and B?

A	Date: 2 May
	Time: 6 hours after midnight

B	Date: 2 May
	Time: 3 hours after midday

______

2. What was the time 30 minutes before?

______

3. 1 hectare = ______ m^2

4. $\frac{1}{3}$ of 24 = $3\overline{)24}$ = $\frac{24}{3}$ = 24 ÷ 3 = ______

5. 1200mL = 1.2L, 2300mL = 2.3L,

   3400mL = ______ L

6. $\frac{4}{10} + \frac{5}{10}$ = ______

7. Draw beads to show the amount 4302.

T/Th | Th | H | T | O

8. Rounding 3.23 to the nearest tenth equals 3.2, so rounding 4.74 to the nearest tenth equals

______.

9. Which angles are:

   (a) obtuse? ______

   (b) acute? ______

10. Which shapes make up a cylinder?

______ and ______

11. ______, 9.5, 9, 8.5, 8

12. $3\overline{)7}$ (2 r 1) = $\frac{7}{3}$ = $2\frac{1}{3}$

    $4\overline{)9}$ (2 r 1) = $\frac{9}{4}$ = ______

13. What is the perimeter of a 100m by 60m rectangular warehouse?

______

14. The product of 4 and 9 is ______.

15. 109 – 10 = ______

Week 12

Week 12

Wednesday

1. There is 10 hours' difference between clocks A and B. For each, write if the time is in the am or pm.

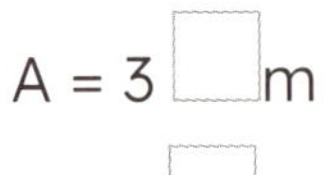
A = 3 ☐m

B = 1 ☐m

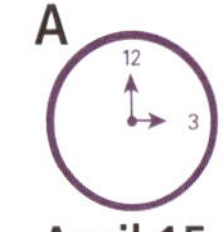

2. $\frac{1}{3}$ of 36 = 36 ÷ 3 = ______

3. 19 – 9 = ______, 190 – 90 = ______

4. What type of triangle is this? ______

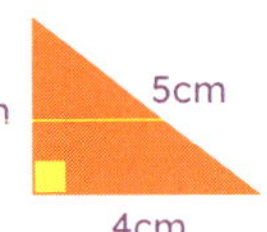

(Not to scale.)

5. 3.0 ÷ 2 = ______

6. What two 2D shapes are faces on a square-based pyramid?

______ and ______

7. Round 6.91 to the nearest whole. ______

8. Rotate a $\frac{1}{2}$ turn.

9. (6 × 4) – 5 = ______

10. What is the area of a 6 by 4 grid?

______ squares

11. What is the length of A?

______cm

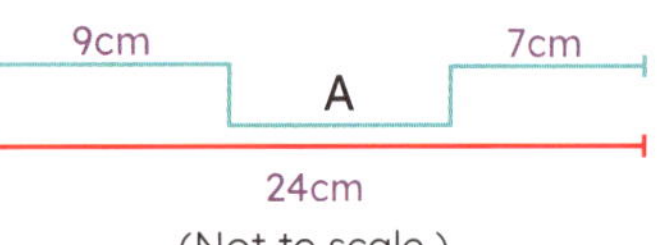

(Not to scale.)

12. 2 + $\frac{4}{10}$ + $\frac{6}{100}$ = ______ . ______

13. What is the date of the extra day in a leap year?

______

14. How many liked cola? ______

Favourite drinks

15. If you earn $11.50 pocket money every week, what amount will you have after six weeks? ______

Thursday

1. 

2. A rectangle has a rotational symmetry to the order of:

☐ 2 ☐ 3

☐ 4

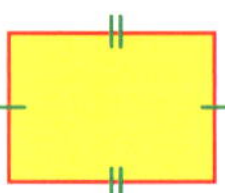

3. 0.7 + 0.4 + 0.2 = ______

4. Which two 2D shapes make up a hexagonal prism?

______ and ______

5. 20 × 19 × 50 = ______

6. 8900mL = ______L

7. Draw a reflection of:

8. Round 6.32 to the nearest tenth. ______

9. 1700 + 4300 = ______

10. When Alicia visited the hairdresser, what was the total cost?

______

Hannah's Hairdressing	
Silky-smooth shampoo	$22.50
Trim	$27.50

11. $\frac{1}{2}$ of 350 = ______

12. 8 × 8 = 32 + ______

13. The probability of a 1 is

______ in ______.

14. $\frac{1}{3} < \frac{1}{2}$ ☐ true ☐ false

15. What is the area of this grid?

______ squares

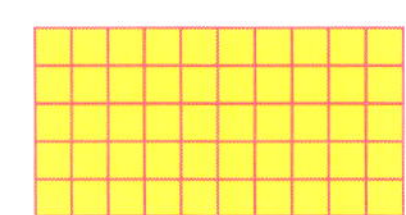

Problem-solving

Alizeh organised a catch-up with Fatima for 12 April at 1400. Fatima misread the catch-up date and time, and arrived at Alizeh's house at 12:00 pm on 14 April.

Let's find out how many hours late Fatima was.

Read the question again. **Think** about the information. Underline the important words.

Tick the strategy you will use to work out the answer:

- estimate and check ☐
- look for patterns ☐
- draw a diagram or picture ☐
- construct a table or graph ☐
- use materials ☐
- act it out ☐
- work backwards ☐
- something else. ☐

Solve it:

Reflect on the question and answer.

Check it. Circle another strategy on the list to work out the answer.

Show it:

Friday Review

1. $\frac{1}{2}$ of 50 = ______

2. Toys 4 me

Bingo	$18.50
Snakes and ladders	$ 5.50
Total	

3. (7 × 4) – 8 = ______

4. $\frac{1}{2} > \frac{1}{10}$

   true ☐ false ☐

5. Round 7.47 to the nearest tenth.

   ______

6. $3 + \frac{4}{10} + \frac{1}{100}$

   = ____ . ____

7. 27 + 25 = ______

8. 5)900 = ______

9. 81, 90, 99, 108,

   ______, ______

10. 4 × 18 × 25

    = ______

11. What was the time half an hour before?

    ______

    2:25

12. In Wednesday's pie graph, how many more liked juice than milk?

    ______

13. Draw a reflection of:

14. There is nine hours' difference between clocks A and B. Write the correct am/pm.

A = 5 ☐m

B = 2 ☐m

15. What is the area of this grid?

    ______ squares

16. What two 2D shapes make up a cylinder?

    ______

    and

    ______

17. What is the length of A?

    ______

    4cm A 3cm

    15cm

    (Not to scale.)

18. Look back at Thursday's spinner. What is the probability of landing on a 5?

    ______

Week 12

Monday

1. The equivalent fraction of $\frac{1}{2}$ is $\frac{\square}{10}$.

2. 250 + 50 + 75 = ________

3. The prime numbers from 2 to 10 are ________.

4. Complete the pattern.

   2, 2.3, 2.6, 2.9, ________

5. Write o for odd or e for even above each number in this magic square.

6	7	2
1	5	9
8	3	4

6. even + even + odd

   = ________

7. odd + odd + odd = ________

8. What is the sum of 6 and 7? ________

9. Use a protractor to draw a 110° angle. Use P as the vertex.

10. $\frac{1}{2} > \frac{1}{6}$ true ☐ false ☐

11. (a) From A, head west 4 squares. Write Y.

    (b) From Y, head directly to (C,4). Write Z.

    (c) What direction is it from Y to Z? ________

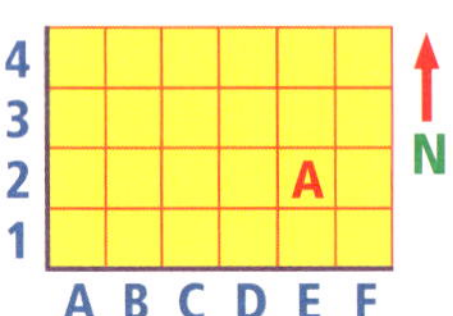

12. The fifth month of the year is ________.

13. What is the length of this pencil?

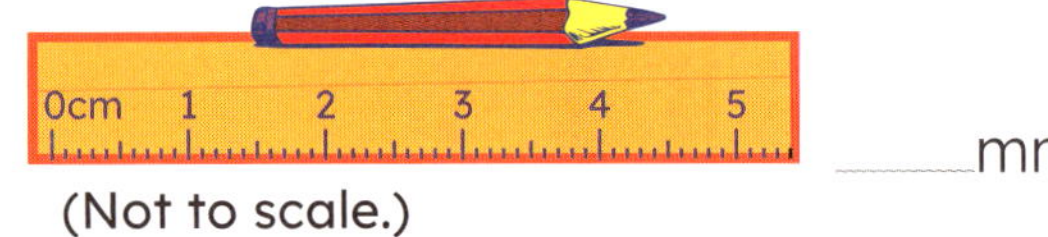

(Not to scale.)

________ mm

14. 10 000 square metres is known as a ________.

15. Round 4.77 to the nearest tenth. ________

Tuesday

1. Write as an equivalent fraction.

   $\frac{1}{2} = \frac{\square}{4} = \frac{\square}{8}$

2. 1t = 1000kg

   1.1t = 1100kg

   1.2t = 1200kg

   1.3t = ________kg

3. Arrange the digits 6, 0, 2, and 7 to create the highest value possible. ________

4. Add 10 to 1990. ________

5. This is the net for a:

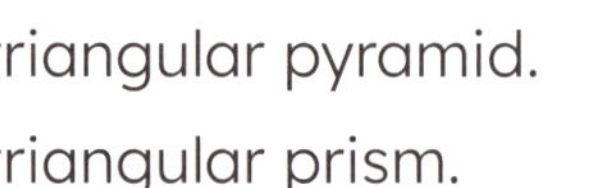

   cube. ☐

   triangular pyramid. ☐

   triangular prism. ☐

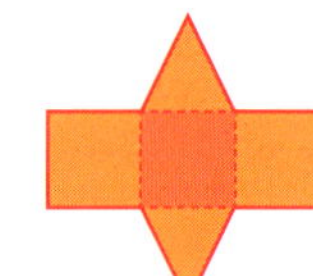

6. 2 × 47 × 50 = ________

7. 140 000 – 90 000 = ________

8. This arrow points to 4mm. Draw an arrow pointing to 18mm.

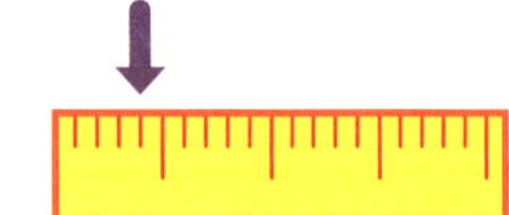

9. Round 6.43 to the nearest tenth. ________

10. What is the date one week earlier than 9 December?

    ____

11. 1000 ÷ 100 = ________

12. 10 × 1.7 = 17

    10 × 2.7 = ________

13. $5 + \frac{7}{10} + \frac{2}{100} =$ ____.____

14. If ∠ ABC = 90°, then ∠ CBD must have a value of ________.

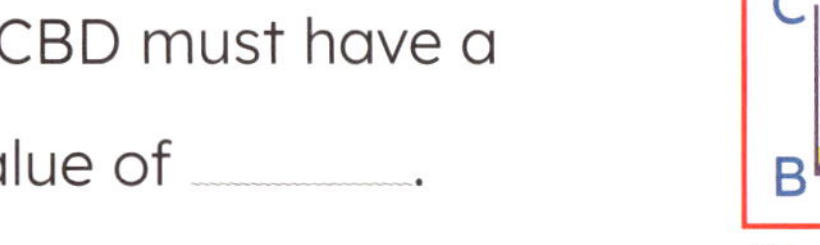

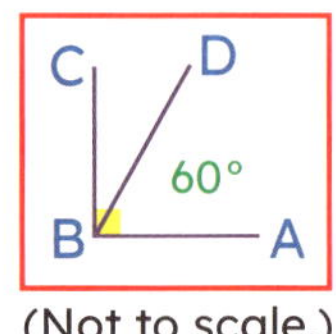

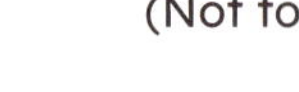
(Not to scale.)

15. Is ∠ CBD an obtuse or an acute angle?

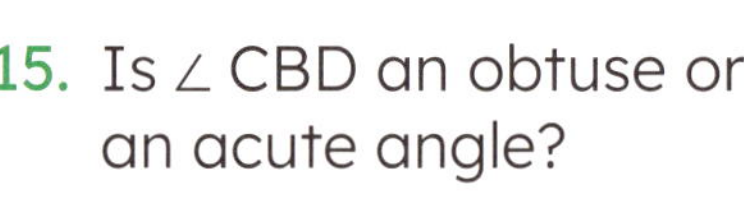

____

Wednesday

1. Write as an equivalent fraction.

   (a) $\frac{2}{3} = \frac{\square}{6}$ (b) $\frac{12}{15} = \frac{\square}{\square}$

2. 150 + 50 + 75 = ________

3. The 24-hour time for 1 pm is:

   0100 ☐ 1300 ☐ 1000 ☐

4. $\frac{1}{4}, \frac{2}{4}, \frac{3}{4}, 1, 1\frac{1}{4}, 1\frac{2}{4}, 1\frac{3}{4},$ ________

5. What is the difference between 145 and 90? ________

6. Which is not a multiple of 3?

18	21	24	27	30	33	36	39
42	45	48	51	53	54	57	60

7. (a) Angles A and B are acute.
   true ☐ false ☐

   (b) Angles A and C are obtuse.
   true ☐ false ☐

   (c) Angles B and D are acute.
   true ☐ false ☐

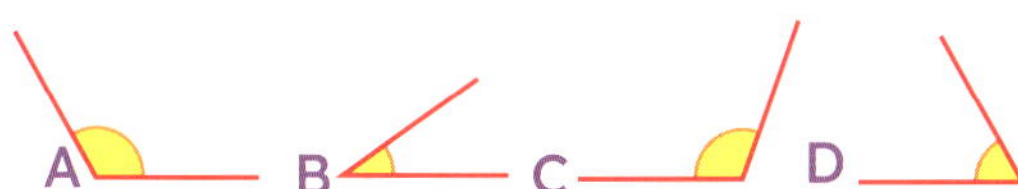

8. A rhombus has a rotational symmetry to the order of:

   2 ☐ 3 ☐ 4 ☐

9. $\frac{8}{10} - \frac{1}{10} =$ ________

10. 63 958 – ________ = 60 000

11. The class teacher receives a $100 budget to spend for 25 students. What is the amount per student?

    $100 ☐ $6 ☐ $4 ☐ $2.50 ☐

12. $6\overline{)402}$ = 201 ÷ 3 = ________

13. This arrow points to 4mm. Draw an arrow pointing to 14mm.

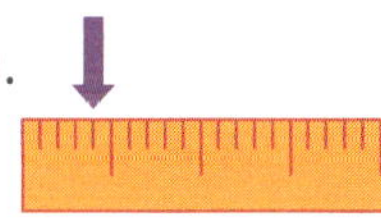

14. Write $\frac{6}{9}$ in its simplest form. ________

15. 1100 – 10 = ________

Thursday

Week 13

1. Write as an equivalent fraction.

   (a) $\frac{4}{6} = \frac{\square}{12}$ (b) $\frac{10}{15} = \frac{\square}{\square}$

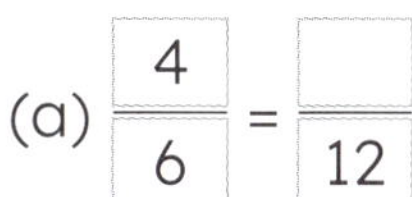

2. Circle the square number.

   45 46 47 48 49 50

3. What date is it four days before 3 April? ________

4. How many edges does a triangular pyramid have? ________

5. What is the time 18 hours after 2 am? ________

6. 900 – 350 = ________

   9000 – 3500 = ________

7. Halve 550. ________

8. 1.1 > 0.9 true ☐ false ☐

9. 1.4 ÷ 2 = ________

10. Use a protractor to draw a 45° angle. Use C as the vertex.

11. Write the next four multiples of 7.

    21, ________, ________, ________, ________

12. Write $5\frac{1}{2}$ as an improper fraction. ________

13. Round 6877 to the nearest hundred. ________

14. What is the probability of the next number rolled being a 2?

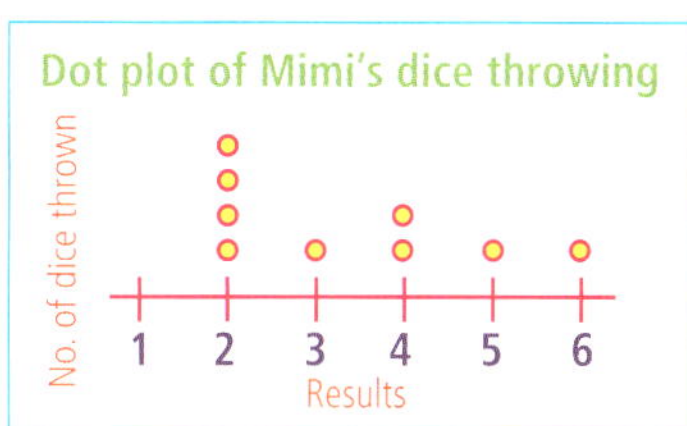

____

15. In a 200g box there were 50g packets of jelly beans. How many 50g packets were there in total?

____

Week 13

Problem-solving

Chef was baking a cake. For the dry ingredients, the recipe called for $\frac{8}{24}$ cups of self-raising flour, $\frac{2}{3}$ cup of cocoa, and $\frac{5}{6}$ cups of coconut.

When Chef went to the cupboard, they noticed all the cup measures were missing except for the $\frac{1}{12}$ cup measure.

Let's find out how many $\frac{1}{12}$ cup measures Chef needed to use in total for this recipe. Write as an improper fraction.

Read the question again. **Think** about the information. Underline the important words.

Tick the strategy you will use to work out the answer:

- estimate and check ☐
- look for patterns ☐
- draw a diagram or picture ☐
- construct a table or graph ☐
- use materials ☐
- act it out ☐
- work backwards ☐
- something else. ☐

Solve it:

Reflect on the question and answer.

Check it. Circle another strategy on the list to work out the answer.

Show it:

Friday Review

1. Write an equivalent fraction.

$\frac{4}{8} = \frac{\square}{4}$

2. Round 16 490 to the nearest thousand.

______

3. 1, 4, 9, 16, ______, 36, 49, ______

4. Write $\frac{10}{12}$ in its simplest form.

______

5. \$20.00 – \$16.20 = ______

6. 10 × 4.39 = 43.9, so 10 × 5.39 = ______

7. The sum of 60 and 70 is ______.

8. (9 + 7) ÷ 2 = 8

true ☐ false ☐

9. 87 239 – ______ = 80 000

10. In a jar of two red and 10 blue marbles, what is the chance of choosing a red?

______

11. How many vertices does a triangular pyramid have?

______

12. Which angle is:

(a) acute? ______

(b) obtuse? ______

13. 2.3t = 2300kg, so

2.4t = ______kg.

14. The arrow points to 4mm. Draw an arrow pointing to 19mm.

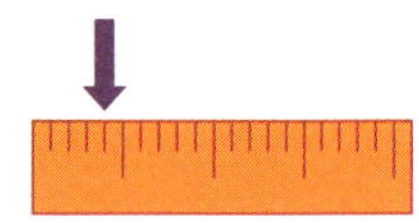

15. The 24-hour time for 4 pm is

______.

16. Draw a net of a cube.

17. How many teachers are aged over 40?

______

18. What is the age difference from the youngest to the oldest teacher?

______ years

Monday

1. Alex has 3 bags with 7 marbles in each bag. Which number sentence describes Alex's total amount of marbles?

   3 + 7 = 10 ☐ 3 × 7 = 21 ☐ 7 – 3 = 4 ☐

2. Rotate a $\frac{3}{4}$ turn clockwise.

3. 54 ÷ 9 = 60 ÷ ______

4. $\frac{15}{4}$ = ______ (mixed number)

   = ______ (decimal)

5. What is the distance from Kingswood to Warrane / Sydney?

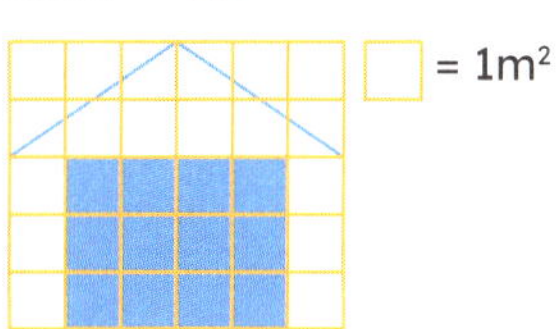

______

6. $\frac{1}{4}$ of 40 = ______

7. 28 ÷ 7 = ______, 4 × ______ = 28

8. The coloured area = ______ m^2

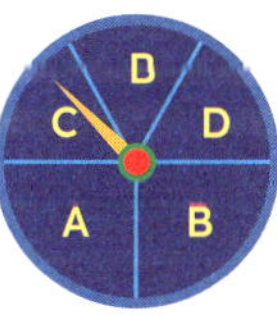

9. Write in descending order.

   $\frac{2}{3}$ $\frac{2}{5}$ $\frac{3}{4}$ $\frac{1}{2}$ $\frac{3}{10}$

   ______ ______ ______ ______ ______

10. 0.8 + 0.7 = ______

11. The probability of a B is:

    $\frac{1}{4} + \frac{1}{6}$ ☐

    $\frac{1}{4} + \frac{1}{3}$ ☐

12. Which state is two hours behind the time of NSW? (Non-daylight saving time.)

    Vic. ☐ Qld. ☐ WA ☐

13. Write the missing factors of 40.

    1, ______, ______, ______, ______, ______, ______, 40

14. Is this time am or pm?

    ______

15. Draw a 60° acute angle. (Use a protractor.)

Tuesday

1. Write this as a number sentence: Angie bought 4 fantasy books at $9 each. What is the total cost?

   ______ × ______ = ______

2. Measure the length of $\overline{XY}$ in mm. ______

   X ———————— Y

3. Circle the number that is not a multiple of 15.

15	30	45	60	75
90	105	125	135	150

4. 300 – 50 = ______

5. $2\overline{)13}$ = ______ r ______

6. The gross mass of a truck was 15t. After unloading its contents, the truck's mass is 8t. What was the net mass of its load? ______ t

7. Does 72 ÷ 7 equal a number greater or less than 10?

   ______

8. Reflect.

9. Round 9.96 to the nearest tenth.

   ______

10. What is the angle size of x?

    ______

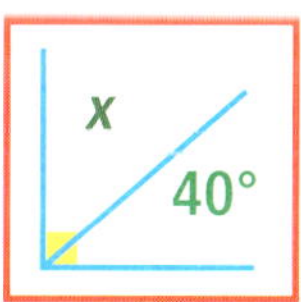

(Not to scale.)

11. Circle the rhombus.

12. How many lots of one dozen can be made from 60 eggs? ______

13. 1cm = 10mm 0.9cm = 9mm

    0.8cm = ______ mm

14. $6 + \frac{1}{10} + \frac{2}{100}$ = ______

15. 5 × [\$50 note] + 3 × [\$20 note] = $ ______

Week 14

Week 14

Wednesday

1. A chef cooked 27 pancakes for 9 students. The pancakes were shared equally. Which equation or number sentence matches how the pancakes were shared?

   27 – 9 = 16 ☐ 27 × 9 = 243 ☐
   27 ÷ 9 = 3 ☐ 27 + 9 = 36 ☐

2. Is this time am or pm? 

   ______

3. 100 – 40 = ______, 1000 – 40 = ______
4. 1.8 ÷ 3 = ___.___
5. Draw a reflection of: 

6. $\frac{1}{3}$ of 36 = ______
7. (a) 2 × 9 = 3 × ______

   (b) 4 × 9 = 6 × ______
8. A square has a rotational symmetry to the order of: 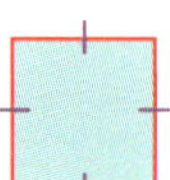

   2 ☐ 3 ☐ 4 ☐
9. If ↑ is north, then ↙ is

   ______________.
10. ______ × 4 = 32, 32 ÷ 8 = ______
11. Write as an equivalent fraction.

    (a) $\frac{5}{10} = \frac{\square}{2}$ (b) $\frac{12}{15} = \frac{\square}{5}$
12. How many have been teaching for exactly 30 years?

    ______

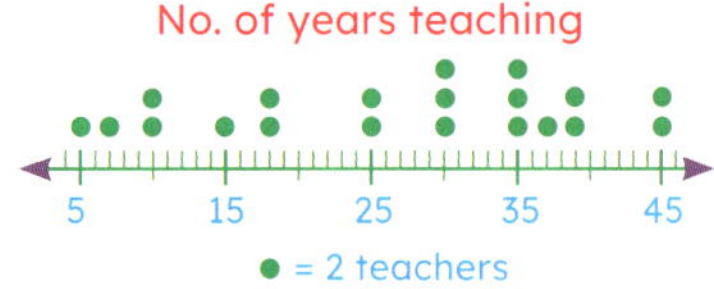

13. \$14.50 + \$2.50 = ______
14. 1.01, 1.02, 1.03, ______, 1.05
15. How many 20c coins make up \$5.60?

    ______

Thursday

1. Write the number sentence for this problem. What is the cost of buying 3 comic books at \$8 each?

   ____
2. Measure the length of $\overline{YZ}$ in mm. ______

   Y|————————|Z
3. 40 ÷ 10 = ______, 400 ÷ 10 = ______,

   4000 ÷ 10 = ______
4. The possible outcomes of flipping two coins are:

   A ☐ HH HT TT TH
   B ☐ HH TH HH TT
   C ☐ TT HT TH TT
5. $\frac{1}{4}$ of 16 = ______
6. 100mm = ______cm
7. Round 6.83 to the nearest tenth.

   ______
8. (4 + 7) × 5 = ______
9. 841 ÷ ______ = 8.41
10. Draw the top view of this 3D object. 
11. 4 × 9 = ______, 40 × 90 = ______
12. Name the type of triangle. 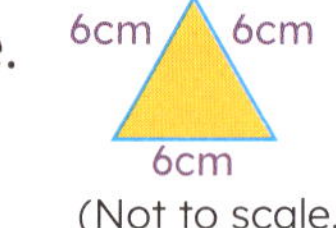

(Not to scale.)

    ____
13. Write one hundred and one thousand as a numeral. ______
14. 64 × 3 = (____ × ____) + (____ × ____)

    = ______ + ______

    = ______
15. What is the cost of buying 5kg of potatoes at 70c per kg?

    ______

Problem-solving

There were 15 people on the bus. At the next bus stop, three people got off and double that number of people got on. How many people were on the bus for the next part of the journey?

Let's write a number sentence for this problem.

Read the question again. **Think** about the information. Underline the important words.

Tick the strategy you will use to work out the answer:

- estimate and check ☐
- look for patterns ☐
- draw a diagram or picture ☐
- construct a table or graph ☐
- use materials ☐
- act it out ☐
- work backwards ☐
- something else. ☐

Solve it:

Reflect on the question and answer.

Check it. Circle another strategy on the list to work out the answer.

Show it:

Friday Review

Week 14

1. ____ × 7 = 42, 42 ÷ 6 = ____

2. Li packed five bananas per bag until all 40 bananas were packed. How many bags were there? (Write as a number sentence.)

3. Is this time am or pm?

4. 1350 + 650 = ____

5. 0.7cm = ____ mm

6. In Wednesday's dot plot, how many more have been teaching for 35 years than five years?

7. $\frac{1}{3}$ of 33 = ____

8. Rotate a $\frac{1}{4}$ turn anticlockwise.

9. 42 × 7
= (____ × 7) +
(____ × 7)
= ____ + ____
= ____

10. 4000 – 10 =

11. What date is it a week after Anzac Day?

12. The time in Tarndanya / Adelaide (SA) is 3 pm (non-daylight saving). What time is it in Warrane / Sydney (NSW)?

13. Round 8.77 to the nearest tenth.

14. 1.10, 1.09, 1.08, 1.07, ____

15. This shape is the base of an object that also has five triangular faces. The object is a:

prism. ☐
pyramid. ☐

16. Measure $\overline{CD}$.

____ cm

C——D

17. A rectangle has rotational symmetry to the order of:

2 ☐ 3 ☐ 4 ☐

18. Using Monday's spinner, what is the chance of an A?

1 in 5 ☐
A is equally as likely as B ☐
more likely than a D ☐

St

Week 15

Monday

1. The guitar was $100. There is 50% off.

It is now ________.

2. (5 × 14) + (5 × 22) = ________

3. How many edges does a cube have? ________

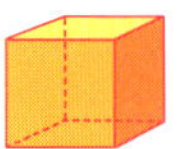

4. $\frac{1}{2} > \frac{1}{20}$ true ☐ false ☐

5. 1km = ________ m

6. Which is the correct pie graph? A cafe recorded 20 drink orders.

7. 4 × 7 = ________

8. Draw the right-side view.

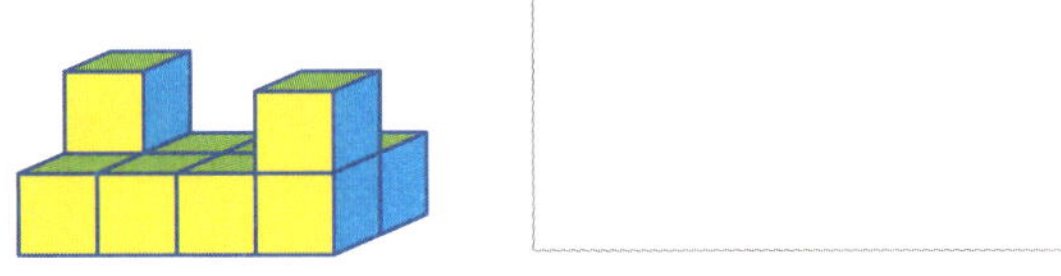

9. $4\overline{)11} = \frac{11}{4} =$ ________ (mixed number)

10. 2.4 ÷ 2 = ________

11. The time in South Australia is 5 pm. The time in Western Australia is ________. (Non-daylight saving time.)

12. 55 + 75 + 35 = ________

13. Complete the multiples of 12.

12	24	36	48
60		84	

14. How many months in one year have 31 days? ________

15. Your weekly pocket money is $12.75. After four weeks, your parents still have not paid. What amount is owed to you?

____

Tuesday

1. The painting was $300, now it costs

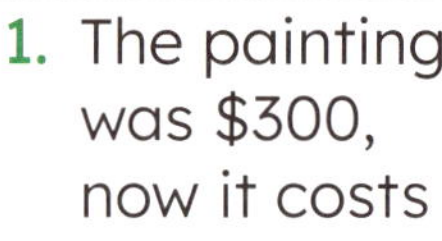

________.

2. $\frac{1}{5}$ of 60. ________

3. Rotate a $\frac{3}{4}$ turn anticlockwise.

4. 320 – 90 = ________

5. Double: 26 = ________ 38 = ________

6. 6 × 60 = 360, so 6 × 59 = ________.

7. 10 – 0.1 = ________

8. This is a

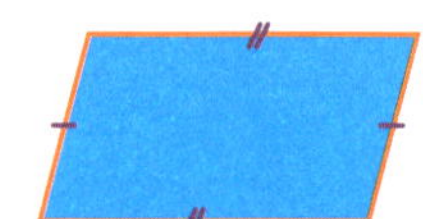

________.

9. Use a protractor. Draw a 65° acute angle. Label the vertex as C and the arms as A and B.

10. Arrange the figures 8, 5, 9, and 0 into the highest value.

____

11. 2ha = ________ m^2

12. What would be the area of a 7 by 10 grid? ________ squares

13. 1000 ÷ 100 = ________

14. $2 + \frac{8}{10} + \frac{4}{100} =$ ____.____

15. Hugo glued the 2D shapes and made a:

sphere. ☐ triangular prism. ☐

cube. ☐ square prism. ☐

Wednesday

1. The boat was $120, now it costs ________.

2. $x + y = 15$
   If $x = 8$, then what is the value of y? ______

3. 2700m = ________km

4. Which set of numbers has 3, 4, and 6 as factors?

   12, 16, 18 ☐ 12, 24, 36 ☐ 12, 24, 32 ☐

5. $\frac{1}{2} + \frac{1}{2} + \frac{1}{2} =$ ________

6. How many laps of a 50-metre pool would have to be swum to reach 1000m?

   ______

7. Draw a square with 2cm sides and colour 25% of it.

8. The fraction at:

   A = ______ B = ______ C = ______

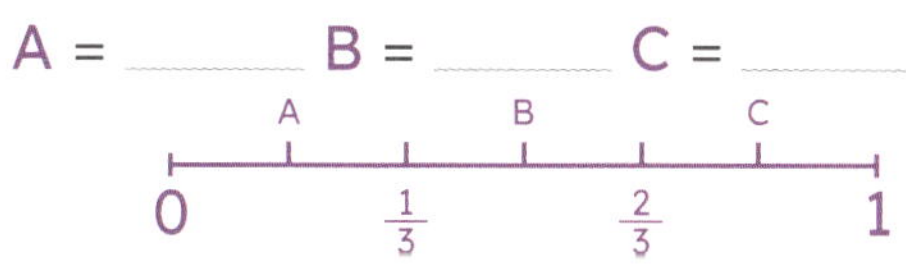

9. 10% discount on $25 is ________.

10. 90 – 11 = ______

11. How many metres of fencing would you need to fence sides A, B, and C of the garden?

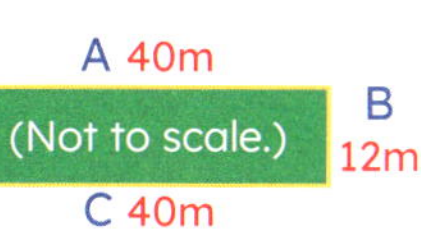

   ______

12. 18 997, 18 998, ________, 19 000

13. Round 4.94 to the nearest whole number. ______

14. (a) 14 × 10 = ______

    (b) 14 × 9 = ______

15. (5 × 24) + (5 × 18) = ________

Thursday

1. The book was $45, now it costs ________.

2. $\frac{1}{5}$ of 120 = ______

3. What would be the length of a fence with measures of 300m on two sides, 450m on one side, and 650m on another?

   ____

4. $15.50 + $4.50 = ________

5. Which set has 2 composite numbers?

   2, 20 ☐ 3, 9 ☐ 9, 15 ☐ 4, 17 ☐

6. An obtuse angle is > 90° and

   < ________°.

7.

8. 4.17 × 10 = ______

9. If ↑ is north, then ↘ is ________________.

10. $\frac{18}{5}$ = ______ (mixed number)

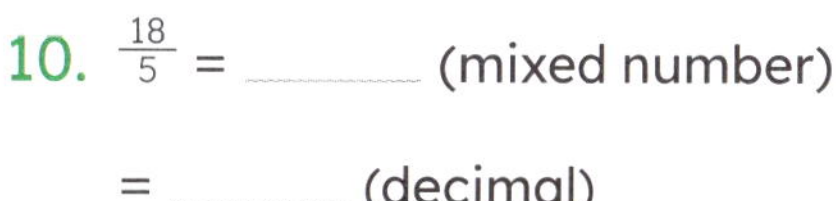

    = ______ (decimal)

11. Add the next five results to the plot graph:

    (3, 4), (2, 2), (5, 1), (1, 1), (6, 3).

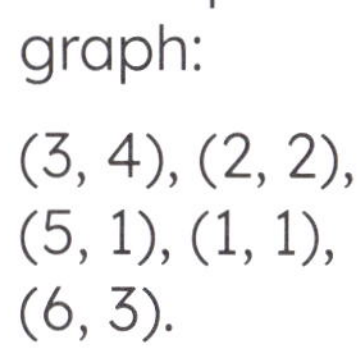

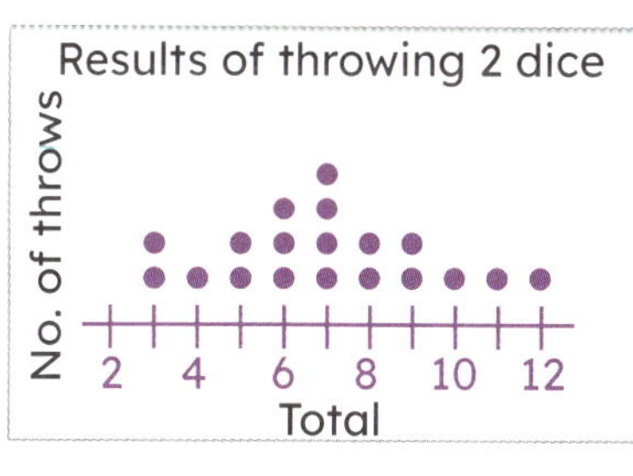

12. Which two numerals are multiples of 3 and 5?

    12 ☐ 24 ☐ 15 ☐ 45 ☐

13. Halve 19 × 10.

    19 × 20 ☐ 19 × 5 ☐ 9.5 × 5 ☐

14. 5000 – 20 = ________

15. Round 10.34 to the nearest tenth. ______

Week 15

Week 15

Problem-solving

At the end of the day, a florist discounts their flower bunches by 5% per flower.

Let's find out how much it would cost to buy these two flower bunches at the discounted prices.

Read the question again. **Think** about the information. Underline the important words.

Tick the strategy you will use to work out the answer:

- estimate and check ☐
- look for patterns ☐
- draw a diagram or picture ☐
- construct a table or graph ☐
- use materials ☐
- act it out ☐
- work backwards ☐
- something else. ☐

Solve it:

Reflect on the question and answer.

Check it. Circle another strategy on the list to work out the answer.

Show it:

Friday Review

1. $\frac{1}{5}$ of 80 = ________
2. 7 × 60 = 420

   7 × 59 = ________
3. 3ha = ________ m^2
4. What is the value of the 3 in 2.73?

   ____
5. $4\overline{)15}$ = $\frac{15}{4}$

   = ________ (mixed number)
6. The book was $20, now it costs ________.

7. 1000 ÷ 100

   = ________
8. $20.00 – $7.50

   = ________
9. Double 745.

   ____
10. 45 + 7 = ________
11. If you swam 1500m in a 50m pool, how many laps did you swim altogether?

    ____
12. 10 000 – 1

    = ________
13. (5 × 15) + (5 × 8)

    = ________
14. If ↑ is north, then ↗ is

    ________.
15. Use a protractor. Draw an obtuse angle of 110°. Label the vertex as P and the arms as Q and R.

16. How many hours' difference?

    ____

17. What is the perimeter of a farm if its rectangular plot is 2km by 1.7km?

    ____
18. A landscaper is planning a yard makeover. There is $100m^2$ of path.

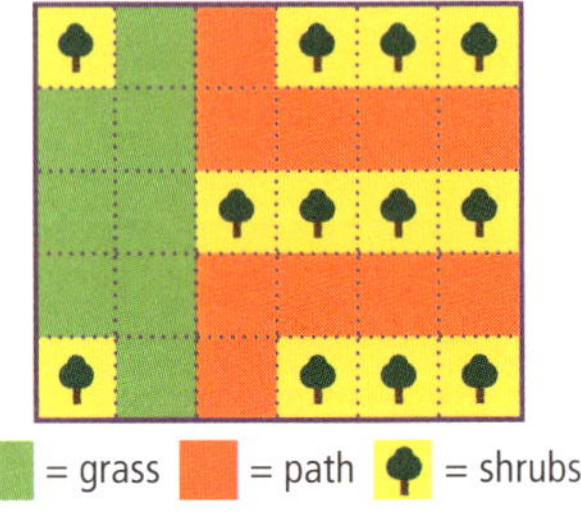

What is the area of the grass?

____

Monday

1. What fraction of students like avocado sushi? ______

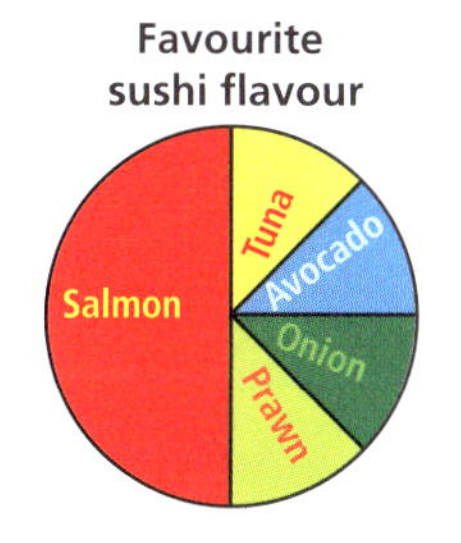

2. What is the least amount of coins needed to make $1.50? ______

3. This object is known as a ______.

4. Complete the number line.

0 [ ] [ ] [ ] 12

5. 46 + 47 = ______, 460 + 470 = ______

6. $5\overline{)125}$ = ______

7. (a) 50 × 9 = ______ (b) 49 × 9 = ______

8. $\frac{1}{4}$ of 100 = ______

9. What is the time in Gulumoerrgin / Darwin (NT) if it is 4 pm in Queensland (non-daylight saving time)?

______

10. 21.09 =

20 + 0.1 + 0.9 ☐ 20 + 1 + 0.9 ☐

20 + 1 + 0.09 ☐ 21 + 0.9 ☐

11. $\frac{1}{2} > \frac{1}{10}$ true ☐ false ☐

12. Measure the length of $\overline{WXY}$. ______mm
Trace over 25% of the line.

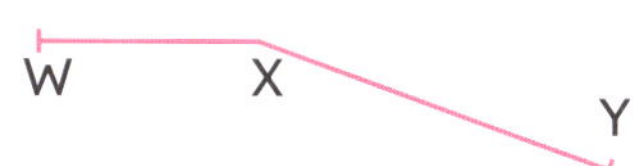

13. Which is heaviest: 1kg, 300g, or 0.1t?

______

14. 0.9 × 100 = ______

15. Double 1945. ______

Tuesday

Week 16

1. Using the pie graph from Monday, what percentage of students like salmon sushi?

______

2. Write the next four multiples of 9.

27, ______, ______, ______, ______

3. What is the least amount of coins needed to make $1.30?

______

4. Tick the quadrilaterals.

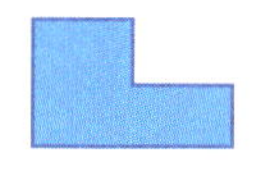 ☐  ☐ 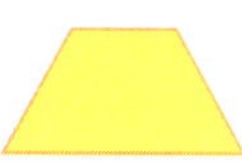 ☐  ☐

5. 0.4 × 100 = ______

6. (28 ÷ 4) × 5 = ______

7. Which equation is correct?

4 × 9 = 38 ☐ 7 × 8 = 56 ☐

6 × 7 = 43 ☐ 4 × 12 = 46 ☐

8.   

9. 6.9m = ______mm

10. 54 ☐ 9 = 6

11. If yesterday was Thursday, what will be the day after tomorrow?

______

12. How many faces does a triangular prism have? ______

13. How many laps would you have to complete in a 50m pool if your goal was to swim 1100m?

______

14. $3 - \frac{5}{8}$ = ______

15. 82 ÷ 100 = ______

Week 16

Wednesday

1. How many chickens laid more than five eggs? ______

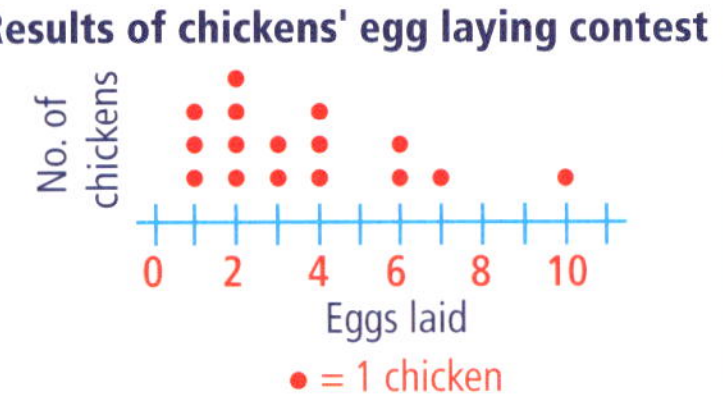

2. 200, 175, 150, ______

3. What is the probability of the spinner landing on D? ______

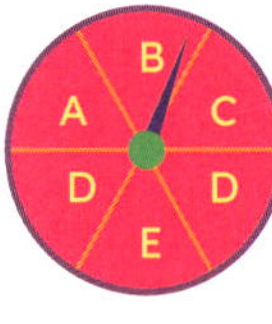

4. 1.27 × 100 = ______

5. Round 9.94 to the nearest tenth. ______

6. 5.5km = ______ m

7. What is the cost of buying 10kg of flour at 65c per kg? ______

8. Round 23 549 to the nearest ten thousand. ______

9. The order of rotational symmetry of a regular pentagon is ______.

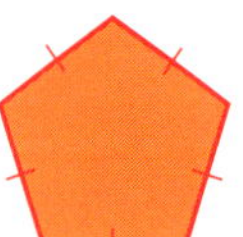

10. How many months are there from 1 March to 1 October? ______

11. 0.7 + 2.3 = ______

12. $\frac{1}{4}$ of 36 = ______

13. Write the missing factors of 24.

    1, 2, 3, ______, ______, ______, 12, 24

14. Colour the numbers that are multiples of 3 and 6.

12	13	14	15	16	17	18
19	20	21	22	23	24	25
26	27	28	29	30	31	32

15. 420 − 60 = ______

Thursday

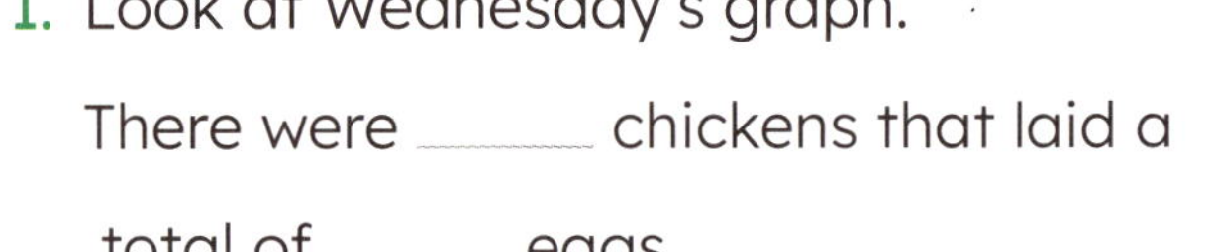

1. Look at Wednesday's graph.

   There were ______ chickens that laid a total of ______ eggs.

2. $\frac{1}{5}$ of 125 = ______

3. If the date is 7 August, what was the date one week ago? ______

4. $5\overline{)205}$ = ______

5. 3.2 ÷ 4 = ______

6. (45 ÷ 9) × (5 + 2) = ______

7. 473 ÷ 100 = ______

8. Which is the longer distance?

   1m ☐ 80cm ☐ 0.1km ☐

9. odd + odd + odd = ______

10. $\frac{2}{3} + 4\frac{1}{3}$ = ______

11. *Fold paper in half.* *Cut shape.* *Unfold.*

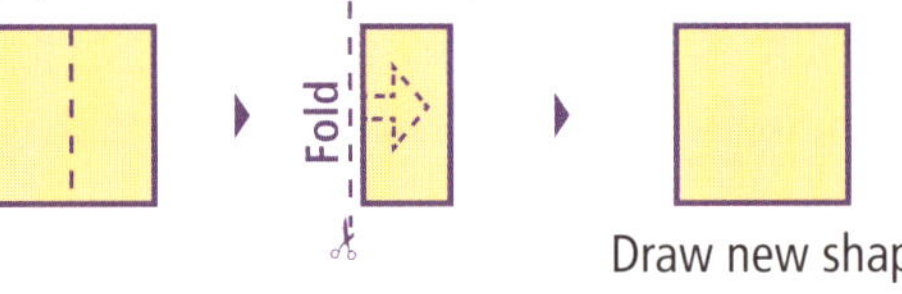

Draw new shape.

12. 150, 300, 450, ______, 750

13. Which equation is true?

    6 × 9 = 54 ☐ 5 × 12 = 70 ☐
    7 × 9 = 56 ☐ 8 × 6 = 46 ☐

14. 10 000 − 100 = ______

15. Which sides are parallel?

    H and C ☐
    H and D ☐
    H and B ☐

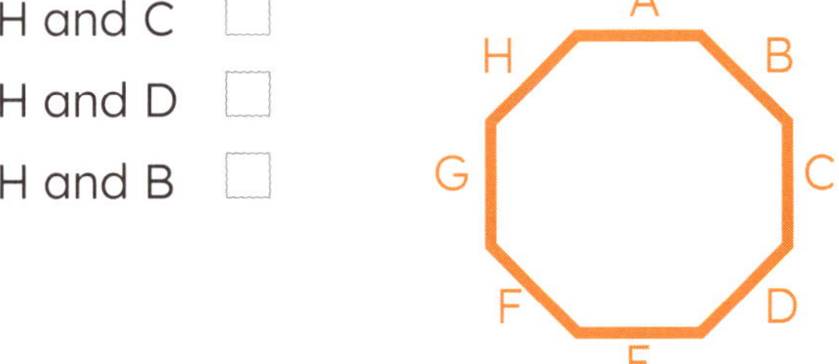

Problem-solving

Class 5 surveyed the shoe sizes of the year groups in their school. Here is the data for Years 4 and 5.

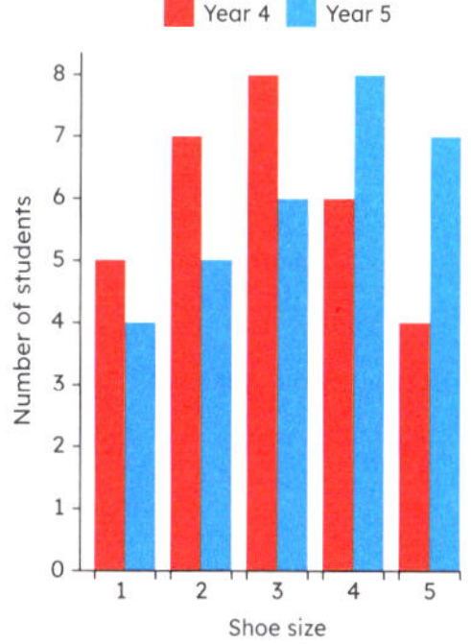

Let's find out what the most common shoe size for both Year 4 and Year 5 was. Based on this information, what might the most common shoe size for Years 6 and 7 be, and why?

Read the question again. **Think** about the information. Underline the important words.

Tick the strategy you will use to work out the answer:

- estimate and check ☐
- look for patterns ☐
- draw a diagram or picture ☐
- construct a table or graph ☐
- use materials ☐
- act it out ☐
- work backwards ☐
- something else. ☐

Solve it:

Reflect on the question and answer.

Check it. Circle another strategy on the list to work out the answer.

Show it:

Friday Review

Week 16

1. What is the least amount of coins needed to make \$1.70?

   ______

2. 160 – 75 = ______

3. Which equation is true?

   $5 \times 9 = 59$ ☐

   $6 \times 9 = 54$ ☐

   $7 \times 7 = 39$ ☐

   $8 \times 7 = 54$ ☐

4. In Wednesday's spinner, what is the probability of landing on A?

   ______

5. $\frac{1}{6} + 3\frac{5}{6} =$ ______

6. $\frac{1}{4}$ of 1000

   = ______

7. 300, 275, 250,

   ______

8. $\frac{4}{5} = \frac{8}{\square}$

9. What is the cost of buying 6kg of almonds at \$11.20 per kg?

   ______

10. 49 ÷ 100 = ______

11. Round 19 590 to the nearest thousand.

    ______

12. What date is one week after 24 May?

    ______

13. A rhombus has rotational symmetry to the order of:

    1 ☐ 2 ☐

    3 ☐ 4 ☐

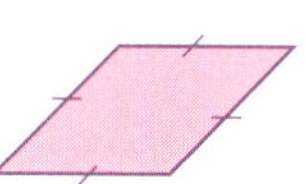

14. 1400mm =

    1.4m ☐

    140m ☐

    0.14m ☐

15. Colour a pair of parallel lines.

16. Colour (D,2), (E,1), (F,3), (A,2), and (B,3).

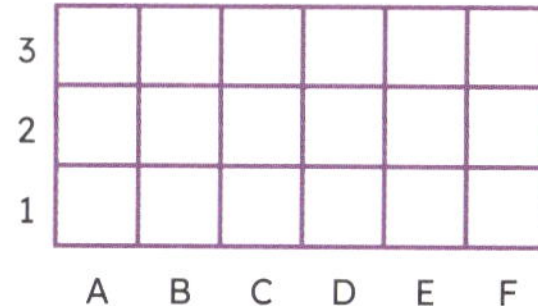

17. Which is heaviest?

    200kg ☐

    2000g ☐

    0.1t ☐

18. What fraction of students liked prawn sushi?

    ______

Favourite sushi flavour

Salmon
Tuna
Avocado
Onion
Prawn

Week 17

Monday

1. What was the time three-quarters of an hour before?

2. The angle between the clock hands is ________°.

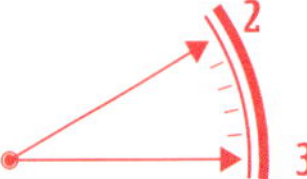

3. 26 + 5 = ________
4. (a) $64 \div 4$ = ________ (b) $640 \div 4$ = ________
5. 3.7km = ________m
6. $3 - \frac{3}{4}$ = ________
7. Write as a number sentence and solve. Po took 16 selfies each day for a week. What is the total number of selfies?
8. This is a ________.

9. 917 241 – ________ = 910 000
10. Halve 850. ________
11. In one year, how many months have 31 days?
12. 63 × 6 = (60 × 6) + (3 × 6)

    = ________ + ________

    = ________

13. 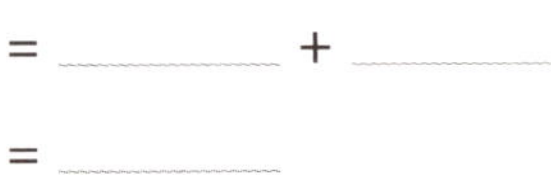 – $19.20 = ________

14. 0.7 + 0.7 = ________
15. What distance would you travel if you left from Swanbourne Beach, passed through Cottesloe Beach, and arrived at Madora Beach?

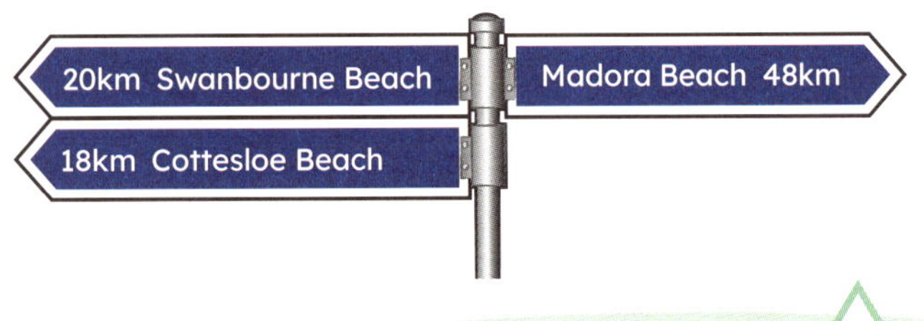

Tuesday

1. What will the time be in 45 minutes?

2. 103 – 6 = ________
3. If Jackson paid $14.80 for their breakfast, what change would they have received from $20?
4. Write question 3 as a number sentence.
5. Colour an arrow to show which way the scales will tip.

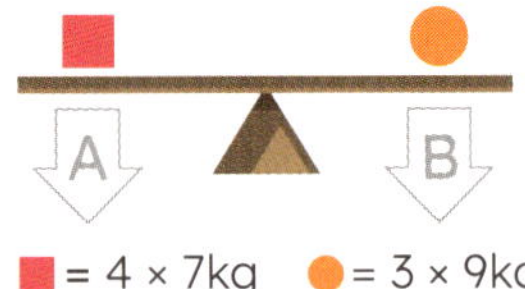

■ = 4 × 7kg ● = 3 × 9kg

6. Write in ascending order.

   0.65 $\frac{1}{4}$ $\frac{5}{10}$ 0.8

7. Round 5.67 to the nearest tenth. ________
8. If ↑ is north, then ↖ is ________.
9. Write $\frac{3}{4}$ as a decimal. ________
10. 4.09 × 100 = ________
11. A chef needs 250mL of beetroot juice. Colour to show 250mL.

12. 300 – 120 = ________
13. Draw three lines of symmetry on the triangle.

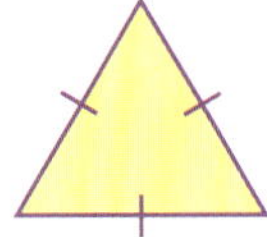

14. Complete the number line.

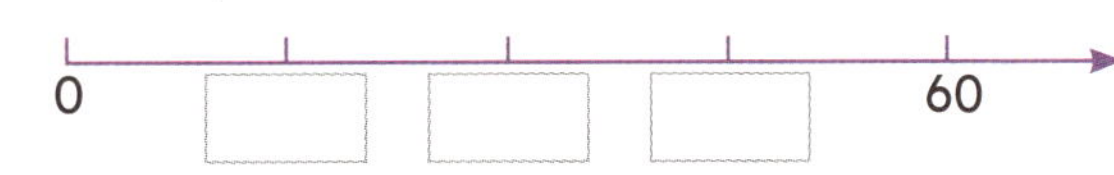

15. 4.5 ÷ 5 = ________

Wednesday

1. What was the time 45 minutes before? 

2. 9094 + 10 = ______

3. $8 - \frac{4}{10} =$ ______

4. The magic square sums to 15 (using numbers 1 to 9) in all directions. Complete the magic square.

8		
	5	
4		2

5. $4\overline{)92}$ = ______

6. 4.6kg = ______ g

7. Write twenty-eight thousand, nine hundred, and fifteen as a numeral.

______

8. 120 – 80 = ______

9. Rotate a $\frac{1}{4}$ turn anticlockwise. 

10. If ↑ is north, then ↙ is

______.

11. The population of Gimuy / Cairns is 169 312. Population means:
    - the number of pops living in a town. ☐
    - the grid coordinates on a map. ☐
    - the number of people living in a town. ☐
    - the number of balloons popped in a year. ☐

12. 376 495 – ______ = 370 000

13. Complete the number line.

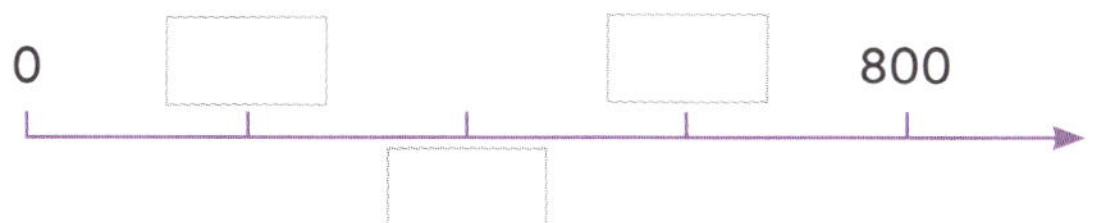

14. 4 – 0.9 = ______

15. If the temperature was 42°C, it would be:
    - extremely hot. ☐
    - warm. ☐
    - freezing. ☐
    - cool. ☐

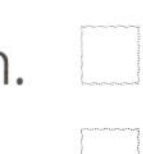

Thursday

Week 17

1. What will the time be in three-quarters of an hour? 

2. Which is the largest amount of money?
    - $\frac{1}{2}$ of \$1000 ☐
    - $\frac{1}{4}$ of \$1750 ☐
    - $\frac{1}{10}$ of \$4800 ☐
    - $\frac{1}{5}$ of \$2000 ☐

3. What is the area of a 6 square by 8 square grid? ______ squares

4. 41 ÷ 6 = ______ r ______

5. 100 000 – 1000 = ______

6. Round $3\frac{3}{7}$ to the nearest whole. ______

7. $\frac{1}{10} > \frac{1}{2}$ true ☐ false ☐

8. How far would you have to travel from the sign to be halfway to Kinjarling / Albany? 

______

9. If a family car weighs about 1.5 tonnes, what would be the total mass of 6 cars?
    156t ☐ 9t ☐ 90t ☐ 1.5t ☐

10. Fill in the missing multiples of 15.

15	30		
			120

11. 9 + 8000 + 70 = ______

12. 2.19 × 100 = ______

13. Double 1.7. ______

14. How many people support the Emus? 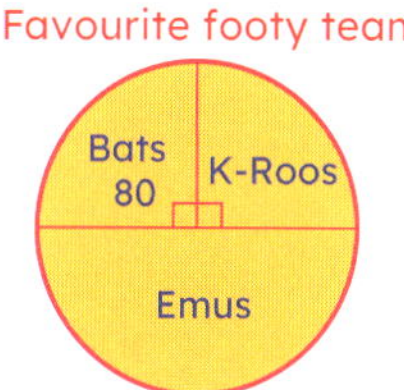

______

15. Every person in town was surveyed. Running into a K-Roos fan in town is:
    - more likely than a Bats fan. ☐
    - less likely than a Bats fan. ☐
    - just as likely as a Bats fan. ☐

Problem-solving

Week 17

Abigail had two flights to catch to get to the holiday destination. This was the boarding pass for their first leg.

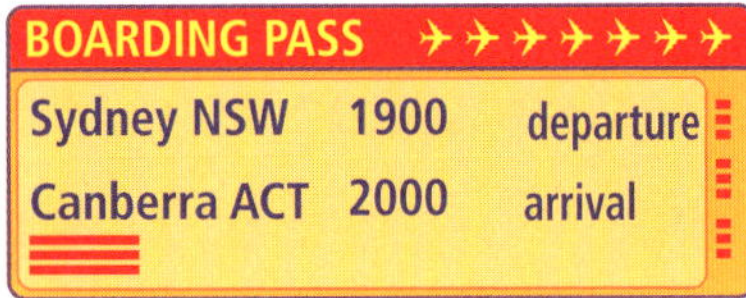

Abigail's first flight was delayed for 50 minutes.

If the stopover between Abigail's two flights was originally planned to be 75 minutes, let's find out how long they had to wait before the second flight departed.

Read the question again. **Think** about the information. Underline the important words.

Tick the strategy you will use to work out the answer:

- estimate and check ☐
- look for patterns ☐
- draw a diagram or picture ☐
- construct a table or graph ☐
- use materials ☐
- act it out ☐
- work backwards ☐
- something else. ☐

Solve it:

Reflect on the question and answer.

Check it. Circle another strategy on the list to work out the answer.

Show it:

Friday Review

1. $\frac{1}{5} < \frac{1}{2}$
   true ☐ false ☐

2. $5\overline{)215}$ = ______

3. \$20.00 – \$18.10
   = ______

4. 63 × 7
   = (60 × 7) + (3 × 7)
   = ______ + ______
   = ______

5. Fill in the missing multiples of 7.

7				
				70

6. If ↑ is north, then ↘ is
   ______.

7. Round 8.87 to the nearest tenth.
   ______

8. 104 + 102 + 5
   = ______

9. What was the time three-quarters of an hour before?
   ______

10. Complete the number line.
    0 ☐ ☐ ☐ 200

11. Write twenty-five thousand and twenty-five as a numeral.
    ______

12. 2.7m = ______cm

13. 462 749 – ______ = 460 000

14. Colour an arrow to show which way the scales will tip.

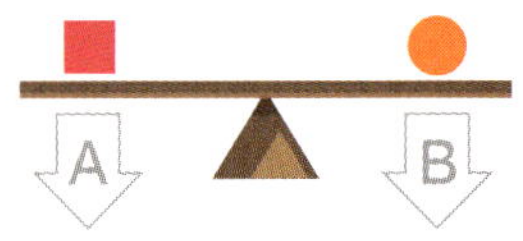

■ = 5 × 9kg ● = 10 × 2.5kg

15. Rotate a $\frac{1}{4}$ (90°) turn to the left.

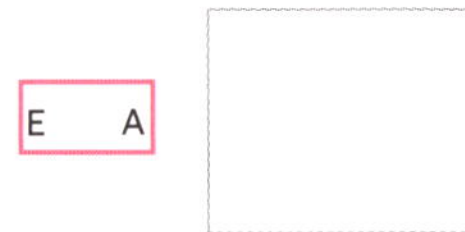

16. If a family car weighs (mass) around 1.5t, what would be the total mass of 4 cars?
    15.4t ☐ 6t ☐
    5t ☐ 4.5t ☐

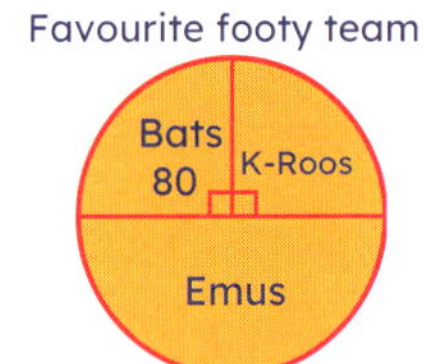

17. How many people were surveyed?
    ______

18. Every person in town was surveyed. Running into an Emus fan in town is:
    more likely than a Bats fan. ☐
    less likely than a Bats fan. ☐

Monday

1. Which four coins are needed to make $1.90?

   ________, ________, ________, and ________

2. What is the perimeter of a rectangular office with dimensions of 25m by 45m?

   ____

3. 1462 ÷ 2 = ________

4. Draw the six lines of symmetry on this regular hexagon.

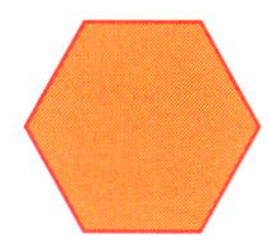

5. Double 900. ________

6. What is the product of 4 and 9? ________

7. What is the value of the 4 in 42 500?

   ____

8. What is the value of a?

   ____

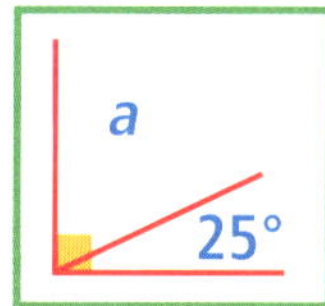

(Not to scale.)

9. 2 × 25 × 3 = ________

10. Half of 17 = 8.5

    Half of 27 = ________

11. Is the letter shape Z symmetrical?

    ____

12. 2.0 – 0.2 = ________

13. 1.7 × 10 = ________

14. A class lined up 25 glue sticks end to end. What was the length?

    ________cm

15. 4, 20, 100, 500, ________

Tuesday

Week 18

1. Which three coins are needed to make $2.30?

   ________, ________, and ________

2. 43 – 8 = ________

3. Would a cone roll in a straight line? ________

4.  

   = $________

5. 990 + 90 = ________

6. What type of triangle is this?

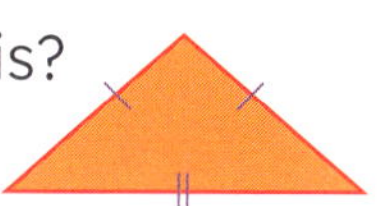

   ____

7. Arrange 4, 0, 7, and 8 to create the largest possible odd number.

   ____

8. What shape is a netball court? ________

9. (5 × 9) ÷ (10 ÷ 2) = ________

10. Measure the length of $\overline{WX}$. ________mm

    W ├────────┤ X

11. What is the cost of 50kg of potatoes at 50c per kg?

    ____

12. If ∠ XOY = 180°, then ∠ XOZ = ________°.

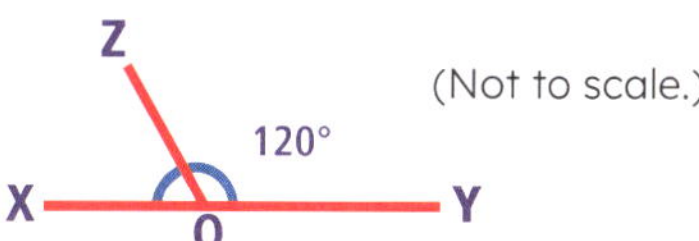

(Not to scale.)

13. 40 + 40 + 40 = ________

14. A coin had a ☆ on one side and a ▲ on the other side. What is the chance of flipping the coin and it landing on a ▲?

    0.5 ☐ $\frac{1}{5}$ ☐ 1 ☐ $\frac{1}{3}$ ☐

15. Write $\frac{1}{4}$ as a decimal. ________

978-1-923005-52-5

Week 18

Wednesday

1. Which five coins are needed to make $3.75?

_____, _____, _____, _____, and _____

2. 2100 – 11 = _____
3. 41 ÷ 10 = _____
4. 3 – 0.3 = _____
5. Halve 39. _____
6. This is a net for a

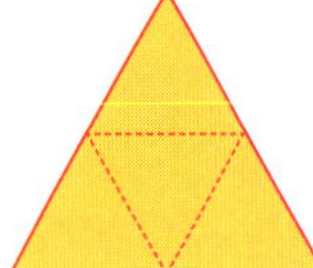

_____ _____.

7. Write the missing factor of 36.

1, 2, 3, 4, _____, 9, 12, 18, 36

8. Colour to show $\frac{7}{4}$. Then write as a mixed fraction.

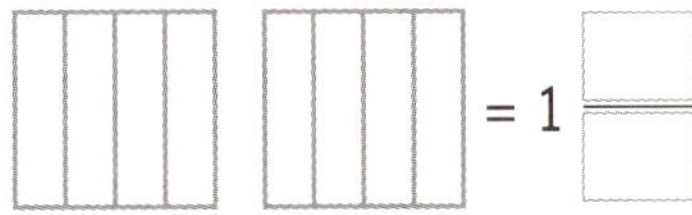

9. 20 × 5 = (20 × 10) ÷ 2 = 100

24 × 5 = (24 × 10) ÷ 2 = _____

10. (6 × 3) ÷ (6 ÷ 2) = _____
11. From A, draw a 45mm line. The line ends at:

_____.

A B C D E F G

12. What is the total cost of buying 100kg of broccoli priced at $2.50 per kg?

_____

13. What is the perimeter of a square with 5cm sides?

_____

14. 2, 6, 18, 54, _____
15. 6 – $\frac{3}{4}$ = _____

Thursday

1. Which four coins are needed to make $4.60?

_____, _____, _____, and _____

2. 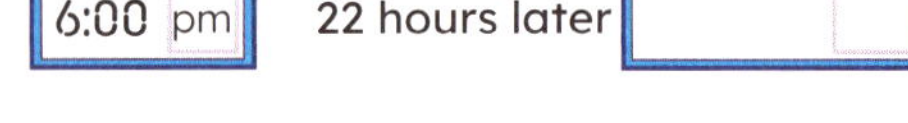

3. The guitar was $600. There is 25% off.

It is now _____.

4. Write 0.008 as a fraction. _____
5. $14.50 + $8.50 = _____
6. $5\overline{)805}$ = _____
7. 9 × 4 = 6 × _____
8. 93 + 28 = _____
9. 9 × 3 = _____
10. 300 000 + _____ = 327 973
11. Draw the right-side view.

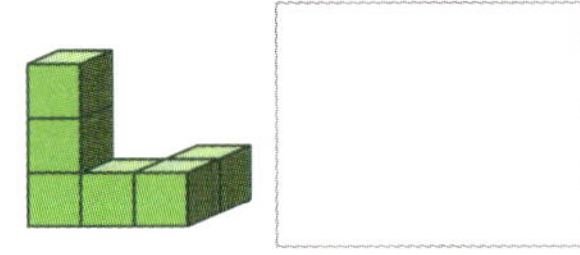

12. Write the number sentence. If you are going for a 50km ride on your bike and so far you have travelled 26km, how far is left to go?

_____

13. Packets of potato chips were boxed.

(a) How many 50g packets will fit in Box A? _____

(b) How many 25g packets will fit in Box B? _____

14. 7 + 100 000 + 4000 + 60 = _____
15. 2.63 × 100 = _____

Problem-solving

Kylie wanted to buy a new soccer ball, priced at $24.95. They checked their money jar and found the following notes and coins: one $10; one $5; three $2; two $1; three 50c; five 20c; two 10c; and one 5c.

Let's find out if Kylie had enough money to buy the ball.

Read the question again. **Think** about the information. Underline the important words.

Tick the strategy you will use to work out the answer:

- estimate and check ☐
- look for patterns ☐
- draw a diagram or picture ☐
- construct a table or graph ☐
- use materials ☐
- act it out ☐
- work backwards ☐
- something else. ☐

Solve it:

Reflect on the question and answer.

Check it. Circle another strategy on the list to work out the answer.

Show it:

Friday Review

Week 18

1. Which four coins can be used to make $3.30?

   ________, ________, ________, ________

2. 110 – 60 = ________

3. $4\frac{8}{8} + \frac{6}{8}$ = ________

4. Halve 31. ________

5. 5, 20, 80, ________, 1280

6. 84 + 29 = ________

7. (a) How many 20g packets will fit in Box A?

   ____

   (b) How many 25g packets will fit in Box B?

   ____

8. (42 × 10) ÷ 2

   = ________

9. 20 + 200 000

   + 5000

   = ________

10. Round 6.47 to the nearest tenth.

    ____

11. Round 24 905 to the nearest thousand.

    ____

12. 3.7 × 10 = ________

13. Write $\frac{1}{4}$ as a decimal.

    ____

14. What size is angle b?

    ____

15. Is this a net for a cube?

    ____

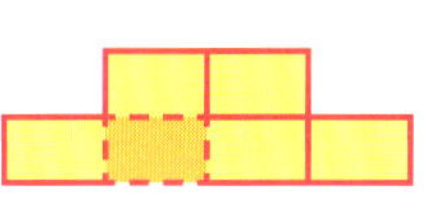

16. Name this triangle.

    ____

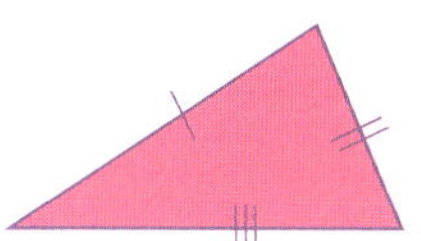

17. What is the perimeter of a rectangular-shaped warehouse that is 45m by 35m?

    ____

18. If a fence costs $20.00 per metre, how much would you pay for 20m of fencing?

    ____

Week 19

Monday

1. Write the next four multiples of 4.

   172, ______, ______, ______, ______

2. Circle the vertex.

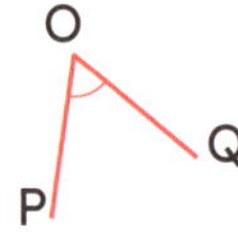

3. What is the value of the 5 in 2.5?

   ______

4. What is the probability of the spinner landing on B?

   ______

5. Complete the number line in decimal form.

   0 ☐ ☐ ☐ 2.0

6. Halve 1. ______

7. Colour 250mL in one colour and 500mL in another.

8. 0.3 > 1   true ☐   false ☐

9. $\frac{1}{4}$ of 24 = ______, so $\frac{1}{2}$ of 24 = ______.

10. This right angle equals:

    90° ☐   100° ☐   180° ☐

11. 1 – 0.9 = ______

12. $x + 20 = y$

    $x = 6$, y = ______

13. (15 – 5) × (7 ÷ 1) = ______

14. Which type of measurement would you use when measuring the size of a tennis court?

    mass ☐   capacity ☐   length ☐

15. 30 + 80 = ______

Tuesday

1. Write the next four multiples of 8.

   88, ______, ______, ______, ______

2. What is the least amount of coins needed to make $7.30?

   ______

3. This is an irregular ______________.

4. 36 ÷ 9 = ______, 4 × ______ = 36

5. 19 × 18 = ______

   20 × 18 = 360

   21 × 18 = ______

6. 9:30 pm $2\frac{1}{2}$ hours later   24-hour time ☐

7. This straight line equals:

   90° ☐   100° ☐   180° ☐

8. 390 ÷ 100 = ______

9. Convert $\frac{15}{10}$ to a decimal. ______

10. When measuring the capacity of a bucket, which unit is best to use?

    kg ☐   L ☐   cm ☐   g ☐

11. What is the area of this grid?

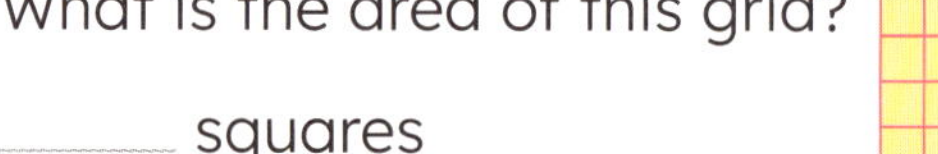

    ______ squares

12. (a) 14 × 0 = ______   (b) 14 × 10 = ______

13. Write twelve thousand and twelve as a numeral.

    ______

14. What is the length of this pencil? ______mm

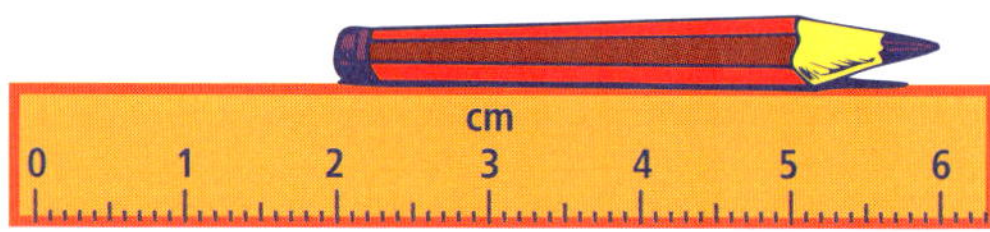

(Not to scale.)

15. (a) $\frac{1}{2}$ of 50 = ______

    (b) $\frac{1}{4}$ of 50 = ______

Wednesday

1. Write the next four multiples of 3.

   45, ______, ______, ______, ______

2. 104 – 7 = ______, 1004 – 7 = ______

3. How many $20 notes are needed to make $1080?

   ______

4. This is an irregular ______________.

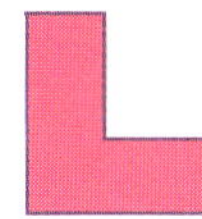

5. 366 days is a:

   school year. ☐ decade. ☐
   leap year. ☐ semester. ☐

6. 60 000 ÷ 60 = ______

7. How many more played netball than tennis?

   ______

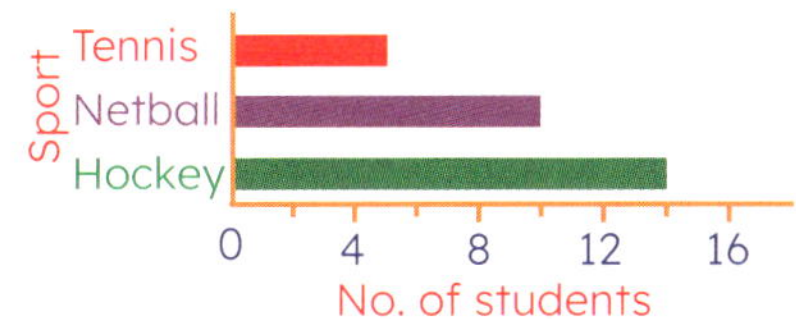

8. $\frac{7}{10}$ = 0.______

9. Jake has five times as many 20c coins as Emma. Jake has $20 of 20c coins. How many 20c coins does Emma have?

   ______

10. 600 + 400 = ______

11. $x \div 5 = y$

    $x = 10$, $y =$ ______

12. 11:30 pm $3\frac{1}{2}$ hours later ______

13. 592 ÷ 10 = ______

14. 11 000, ______, 10 998, 10 997

15. $\frac{21}{4}$ = $4\frac{1}{4}$ ☐ $5\frac{1}{4}$ ☐ $6\frac{1}{4}$ ☐

Thursday

Week 19

1. Write the next four multiples of 5.

   85, ______, ______, ______, ______

2. Colour $\frac{4}{10}$.

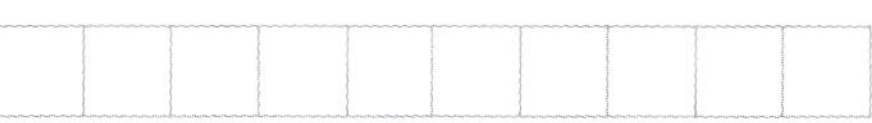

3. 50 × 60 = ______

4. 3000 ÷ 30 = ______

5. 4 – 0.3 = ______

6. Which equation is true?

   4 × 9 = 38 ☐ 7 × 7 = 49 ☐
   8 × 7 = 54 ☐ 4 × 11 = 41 ☐

7. An equilateral triangle has rotational symmetry to the order of three. What is the angle of each rotation?

   60° ☐ 120° ☐ 180° ☐

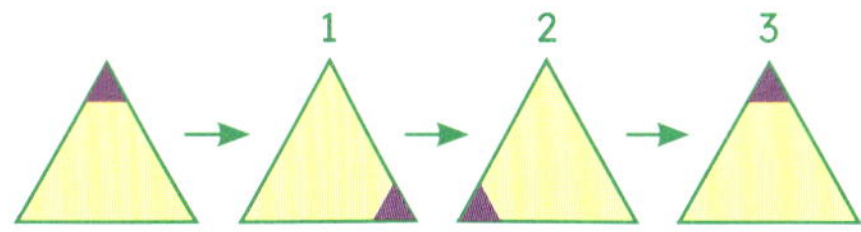

8. 3 + 9 = ______, 0.3 + 0.9 = ______

9. How many weeks are in one year? ______

10. 6 × 3 = 2 × ______

11. What is the least amount of coins needed to make $13.40?

    ______

12. $\frac{3}{10} + \frac{4}{10} =$ ______

13. What number is after 199 999?

    ______

14. The quotient when 45 is divided by 5 = ______ each

15. Match each name to a shape.

    rhombus
    kite
    trapezium

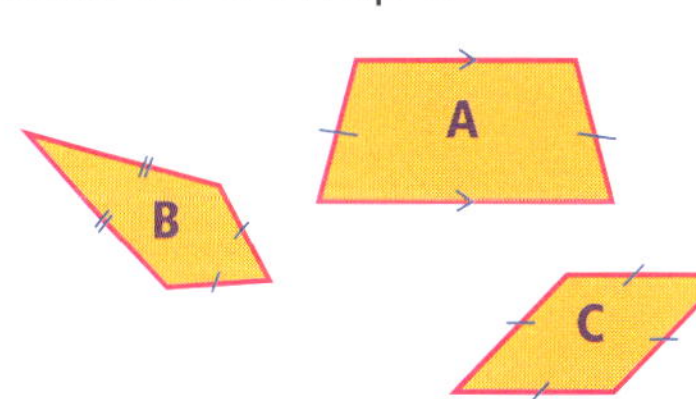

Week 19

Problem-solving

Mia and Talia were sorting stones into piles, to use during a yangamini tournament. Each pile had six stones in it.

Let's find out how many stones they had altogether if they sorted six piles, eight piles, 10 piles, and 12 piles.

Read the question again. **Think** about the information. Underline the important words.

Tick the strategy you will use to work out the answer:

- estimate and check ☐
- look for patterns ☐
- draw a diagram or picture ☐
- construct a table or graph ☐
- use materials ☐
- act it out ☐
- work backwards ☐
- something else. ☐

Solve it:

Reflect on the question and answer.

Check it. Circle another strategy on the list to work out the answer.

Show it:

Friday Review

1. Write the next four multiples of 6.

   150, ____, ____, ____, ____

2. In Wednesday's graph, how many students played either hockey or netball?

   ____

3. 102 – 7 = ____

4. What number is after 110 999?

   ____

5. 72 ÷ 9 = ____

   8 × ____ = 72

6. Halve $\frac{1}{2}$. ____

7. $y \div x = 10$

   $x = 9$, $y =$ ____

8. (11 – 5) × (100 ÷ 10) = ____

9. How many $20 notes are needed to make $2000?

   ____

10. 19 × 19 = ____

    20 × 19 = 380

    21 × 19 = ____

11. 0.7 + 0.4 = ____

12. Convert $\frac{17}{10}$ to a decimal.

    ____

13. 3 × 9 = ____

14. This is an irregular ____.

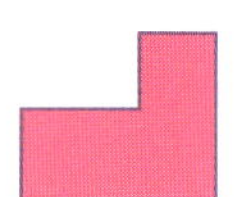

15. Colour a trapezium shape inside the hexagon.

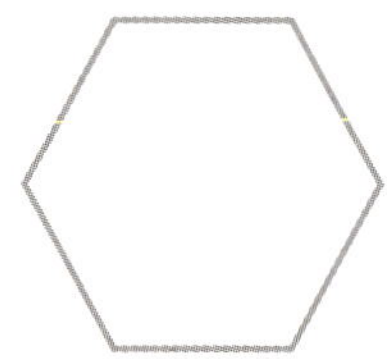

16. 9:30 am

    $3\frac{1}{2}$ hours later

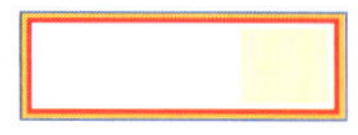

17. What is the area of this grid?

    ____ squares

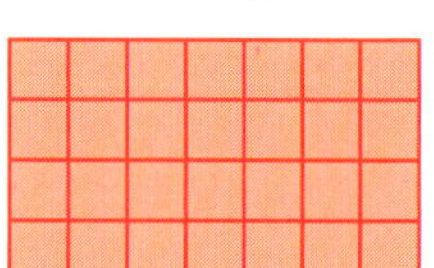

18. What is the probability of the spinner landing on 1?

    ____

Monday

1. Order from least to greatest in value.

   2220 2020 2202 2002 2022

   ______ ______ ______ ______ ______

2. 11:30 pm $3\frac{1}{2}$ hours later ______

3. Write the next four multiples of 6.

   96, ______, ______, ______, ______

4. (7 + 4) + (30 ÷ 3) = ______

5. How many odd-numbered houses are in New Wave Avenue, from 29–45?

   ______

6. How many students were surveyed?

   ______

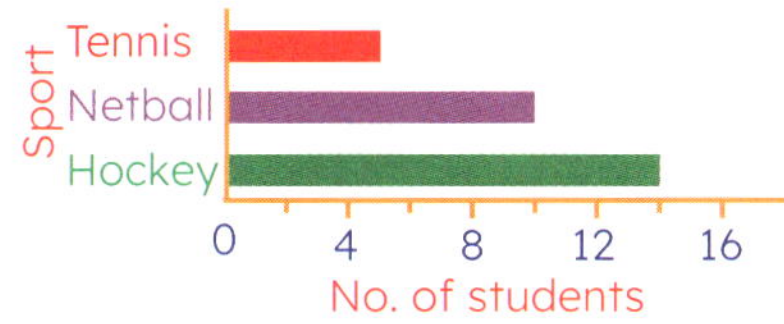

7. A circle = ______ °

8. A clock hand moved from 1 to 3.

   It rotated ______.

9. $\frac{3}{1000}$ = 0.______

10. Complete the number line.

11. Write $\frac{15}{10}$ as a decimal. ______

12. Double 150. ______

13. 1 > 1.1 true ☐ false ☐

14. When measuring mass, which two units are best to use?

    km ☐ kg ☐ g ☐ cm ☐ mL ☐

15. Write three million as a numeral.

    ______

Tuesday

Week 20

1. Order from greatest to least in value.

   7009 7090 4998 7019

   ______ ______ ______ ______

2. 350 + 650 = ______

3. What is the value of the 4 in 2.24? ______

4. A square is ______-dimensional.

5. A cube is ______-dimensional.

6. 0.1 × 5 = ______

7. 4.2 ÷ 6 = ______

8. When measuring capacity, which two units are best to use?

   km ☐ kg ☐ g ☐ L ☐ mL ☐

9. Packets of potato chips were boxed.

   (a) How many 20g packets will fit in Box A? ______

   (b) How many 25g packets will fit in Box B? ______

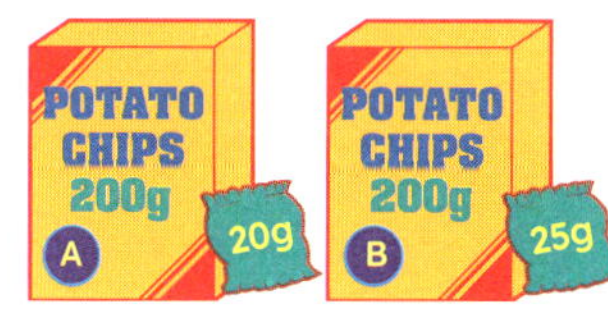

10. 15, ______, 45, 60, 75

11. $y \times x = 27$ $x = 9, y =$ ______

12. 70 × 3 = 210, 69 × 3 = ______,

    68 × 3 = 204

13. 13 – 8 = ______, 1300 – 800 = ______

14. Surfie Sam has some unusual surfboards. They need to wax each edge of all the surfboards. How many edges need waxing? ______

15. To look cool, Surfie Sam paints a blue line on each surfboard's line of symmetry. Draw these lines.

Wednesday

1. Write in ascending order.

2012 2110 2010 2100

_____ _____ _____ _____

2. $3 + \frac{4}{10} + \frac{9}{10} =$ ____.____

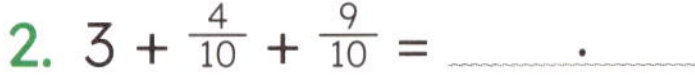

3. What is the value of 9 in 3.69? _____

4. How far did the car travel from A to B? _____

5. 1.7km = _____ m

6. $a \times b = 20$

$a = 4$, $b =$ _____

7. 40 000m^2 = _____ ha

8. 29 × 14 = _____

30 × 14 = 420

31 × 14 = _____

9. Complete the times (non-daylight savings).

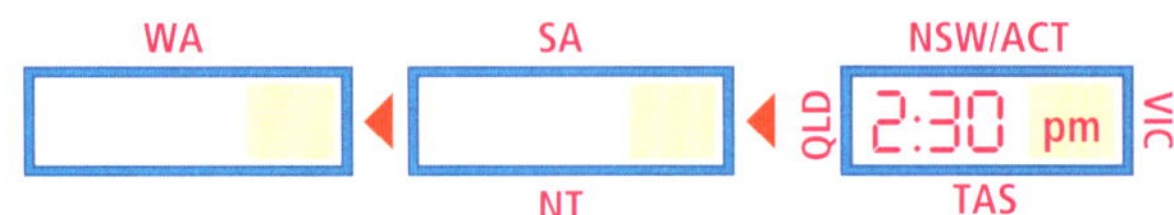

10. Double $19\frac{1}{2}$. _____

11. A straight line = _____ °

12. 25, _____, 75, 100, 125

13. (a) The outcomes are: _____

(b) The probability of a 1 is _____.

14. 104 – 8 = _____, 1004 – 8 = _____

15. $\frac{1}{5}$ of 10 = 2, $\frac{1}{5}$ of 20 = 4, $\frac{1}{5}$ of 30 = _____

Thursday

1. Write in descending order.

4980 4890 4089 4098

_____ _____ _____ _____

2. 650 + 650 = _____

3. 07:00 pm $4\frac{1}{2}$ hours later [24-hour time] _____

4. 900 – 10 – 5 = _____

5. 104 – 6 = _____, 1004 – 6 = _____

6. 3.1kg = _____ g

7. 0.9 + 0.4 = _____

8. $\frac{9}{100} + \frac{3}{10} =$ 0._____

9. What 2D shape would you see in the cross-section of this 3D object?

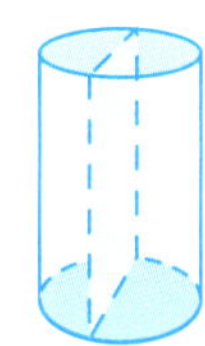

_____

10. Halve 47. _____

11. The four angles at the intersection of the diagonals are:

90° ☐
108° ☐
0° ☐
180° ☐

12. 40, _____, 120, 160, _____, _____, _____

13. 29.6 ÷ 100 = _____

14. On a hot day at a rock concert, 300L of water was sold. The capacity of each water bottle was 600mL. How many water bottles were sold?

_____

15. The book was $15. It now costs _____.

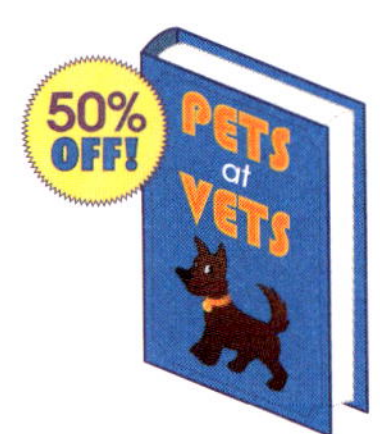

Problem-solving

Five friends started reading the same 120-page book at the same time. After the first day, Desi had read 0.2 of the book, Bernardo $\frac{1}{4}$, Mika $\frac{6}{8}$, Pranay 0.6, and Gillian $\frac{1}{3}$.

Let's find out how many pages each friend has left to read. Order names from least to most pages left to read.

Read the question again. **Think** about the information. Underline the important words.

Tick the strategy you will use to work out the answer:

- estimate and check ☐
- look for patterns ☐
- draw a diagram or picture ☐
- construct a table or graph ☐
- use materials ☐
- act it out ☐
- work backwards ☐
- something else. ☐

Solve it:

Reflect on the question and answer.

Check it. Circle another strategy on the list to work out the answer.

Show it:

Friday Review

Week 20

1. 650 + 750 = ______
2. 102 – 9 = ______
   1002 – 9 = ______
3. The headphones were $500.

They now cost ______.

4. 35, ______, 105, 140
5. 1.3m = ______cm
6. What is the value of the 3 in 2.23? ______
7. Write the numbers in ascending order.
   4430 4040 4340
   4304 4300 4403
   ______ ______
   ______ ______
   ______ ______
8. $\frac{7}{1000}$ = 0.______
9. Double $17\frac{1}{2}$. ______
10. $\frac{8}{100} + \frac{6}{10}$ = 0.______
11. (6 + 7) + (40 ÷ 5) = ______
12. How many even numbers are from 1 to 16? ______
13. 1.5 ÷ 3 = ______
14. 0.1 × 9 = ______
15. A straight line is ______°.
16. 13:30
    $5\frac{1}{2}$ hours before
    ☐ (24-hour time)
17. A triangular pyramid has faces numbered 1, 2, 3, and 4. When rolled, what is the chance of it landing on 3?
    - 1 in 3 ☐
    - $\frac{1}{4}$ ☐
    - $\frac{1}{3}$ ☐
    - $\frac{3}{4}$ ☐
18. Look back at Monday's graph, the difference between the most popular and the least popular sport is ______.

Monday

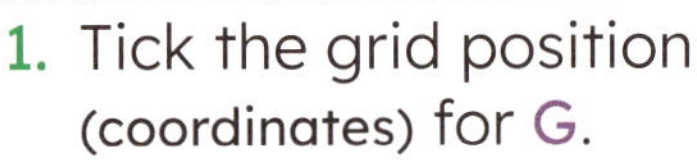

1. Tick the grid position (coordinates) for G.

   (1,2) ☐ (2,1) ☐

   (2,2) ☐ (1,4) ☐

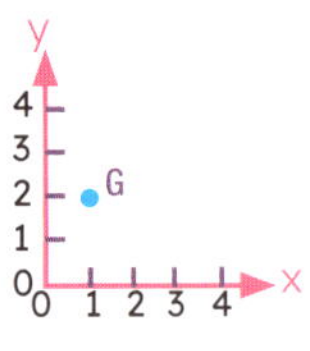

2. 204 – 7 = ________

3. This jug has a capacity of 2L. Each mark equals:

   100mL. ☐ 250mL. ☐

   500mL. ☐

4. Order from the greatest to the least value.

   0.9 1 0.5 0.1

   ________ ________ ________ ________

5. 37 589 + ________ = 237 589

6. $\frac{142}{100}$ = ________ (decimal) = ________%

7. 11:45 pm 20 minutes later ________

8. (a) 4 × 6 = ________ (b) 8 × 6 = ________

9. Write four million and twelve thousand as a numeral.

   ____

10. 19.09 × ________ = 190.9

11. The factors of 12 are:

    1, ________, ________, ________, ________, 12

12. Round 7.4 to the nearest whole number.

    ____

13. What is the probability of all your class wearing school uniform? Place a dot on the number line.

14. Which equation is true?

    72 ÷ 8 = 11r2 ☐ 60 ÷ 10 = 6 ☐

    49 ÷ 7 = 6 ☐ 32 ÷ 4 = 7 ☐

15. $\frac{1}{5} + \frac{3}{5}$ = ________

Tuesday

1. Write the correct coordinates for B.

   ____

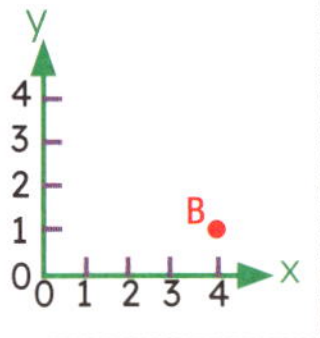

2. 2000 – 250 = ________

3. 99 + 7 = ________

   999 + 7 = ________

4. 43 ÷ 5 = ________ (mixed number)

   = ________ (decimal)

5. (a) 9 × 0 = ________ (b) 9 × 10 = ________

6. What is the distance from:

   (a) D to B? ________

   (b) A to C? ________

   (c) B to C? ________

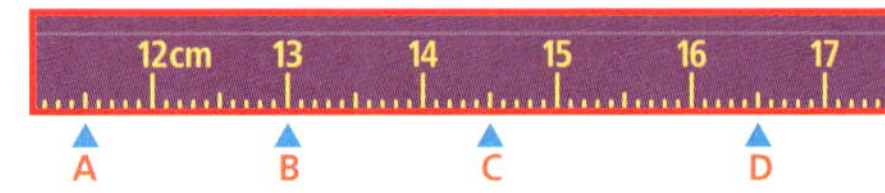

7. What is the perimeter of a 7 by 5 grid?

   ____

8. What is the area of a 7 by 5 grid?

   ________ squares

9. 10:15 am 20 minutes later ________

10. $\frac{3}{4}$ = ________%

11. The highest common factor (HCF) of 12 and 16 is:

    2. ☐ 4. ☐ 6. ☐ 8. ☐

12. 10 × $2.75 = ________

13. $\frac{1}{4}$ of 120 = ________

14. Halve 45. ________

15. Draw a $\frac{1}{4}$ turn clockwise.

Wednesday

1. Plot A at coordinate (3,4).

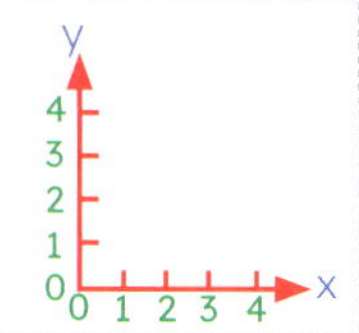

2. 99 + 9 = ________, 999 + 9 = ________

3. Mrs Money paid $485 000. How much did Mrs Money save?

____

4. Double 195. ________

5. 15 × 8 = 15 × 2 × 2 × 2 = ________

6. 15 – 6 = ________

7. 692 ÷ 10 = ________

8. *Fold paper in half.* *Cut shape.* *Unfold.*

Fold

Draw new shape.

9. $6\overline{)21} = \frac{21}{6}$ = ________ (mixed number)

10. 83 × 4 = (______ × ______) + (______ × ______)

= ________ + ________

= ________

11. $20.00 – $3.60 = ________

12. $\frac{1}{3}$ of 36 = ________

$\frac{2}{3}$ of 36 = ________

13. Angles A and D are:

obtuse. ☐ acute. ☐

B C A D

14. $\frac{186}{100}$ = ________ (decimal) = ________%

15. What is the probability of you going to the shops this afternoon? Place a dot on the number line.

Thursday

1. Plot A at coordinate (4,2) and B at (1,3).

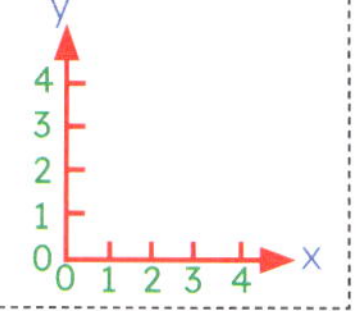

2. 5000 – 350 = ________

3. 119 + 6 = ________

4. Write in ascending order.

0.7 0.4 1.0 0.8 1.2

________ ________ ________ ________ ________

5. A large rectangular cattle station's dimensions were 2km by 14km.

What is the area? ________

6. (a) How many liked limes? ________

(b) What percentage liked apples?

____

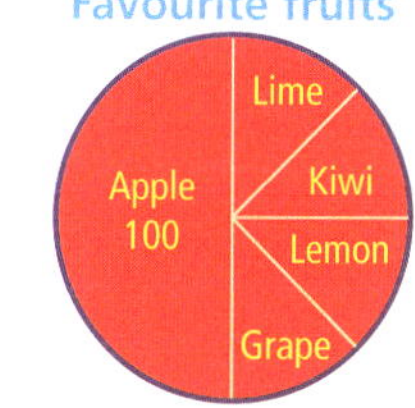

7. 0.9 × 3 = ________

8. (50 ÷ 2) – (10 ÷ 2) = ________

9. 70, 140, 210, ________, 350

10. 23mm = ________cm = ________mm

11. 2 – 0.1 = ________

12. The time difference is ________.

13. $4\frac{1}{2} + 3\frac{1}{2}$ = ________

14. The highest common factor of 18 and 24 is:

2. ☐ 4. ☐ 6. ☐ 9. ☐

15. How far is it from Warrane / Sydney to Newcastle?

________km

Week 21

Problem-solving

Week 21

Keira drew a coordinate grid like the one used for Thursday's question 1. Keira wanted to draw a pentagon on the grid.

Let's find out what coordinates could be used to draw the pentagon.

Read the question again. **Think** about the information. Underline the important words.

Tick the strategy you will use to work out the answer:

- estimate and check ☐
- look for patterns ☐
- draw a diagram or picture ☐
- construct a table or graph ☐
- use materials ☐
- act it out ☐
- work backwards ☐
- something else. ☐

Solve it:

Reflect on the question and answer.

Check it. Circle another strategy on the list to work out the answer.

Show it:

Friday Review

1. Draw a $\frac{3}{4}$ turn clockwise.

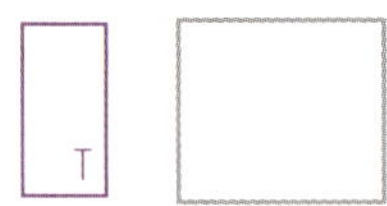

2. 8295 + ________ = 178 295

3. $1.83 × 10 = ________

4. 16 – 7 = ________

5. The highest common factor (HCF) of 27 and 36 is:
   3. ☐ 6. ☐
   9. ☐ 12. ☐

6. 16 × 8 = 16 × 2 × 2 × 2 = ________

7. 99 + 6 = ________
   999 + 6 = ________

8. 0.4 × 2 = ________

9. 80, 160, ________, 320, 400

10. $\frac{1}{3}$ of 30 = ________
    $\frac{2}{3}$ of 30 = ________

11. $7\frac{1}{2} + 8\frac{1}{2} + 2\frac{1}{2}$ = ________

12. What is the distance from Wallup to Horsham?

____

17km Horsham | 23km Wallup

13. Plot A at coordinate (3,2).

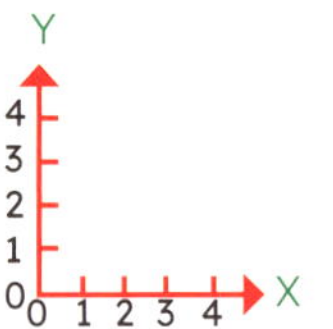

14. The quotient when 60 is divided by 10 is ________.

15. How many 15g packets will fit in Box A?

____

16. How many 30g packets will fit in Box B?

____

17. What is the probability of you watching TV after school? Place a dot on the number line.

impossible ——— certain

18. How many people liked apples, lemons, or kiwis?

____

Favourite fruits

St P

Monday

1. (24 × 5) + (18 × 5) = ______

2. What is the time if the long hand is on 7 and the short hand is between 7 and 8?

______

3. Share $20.00 equally among 8 children.

______

4. 

= $ ______

5. 1999 + 7 = ______

6. What is the place value of 3 in 2.973?

- 0.1 (tenths) ☐
- 0.01 (hundredths) ☐
- 0.001 (thousandths) ☐

7. Add one bead to each place value and write the new amount.

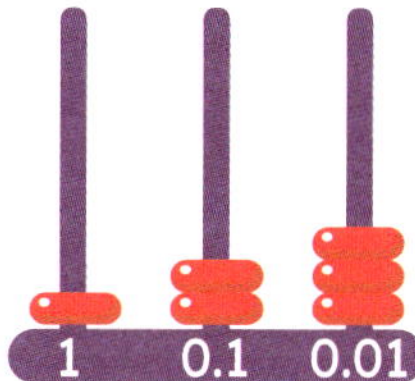

______

8. If a bus is due to arrive at 6:15 pm and your digital watch displays 5:55 pm, how many minutes are left before it is due? ______

9. 10 000, ______, 6000, 4000, 2000

10. This is an irregular ______.

11. The sum of all the angles in a triangle is ______.

12. 1.5m = ______ mm

13. Draw a 60° angle (use a protractor). Make O the vertex.

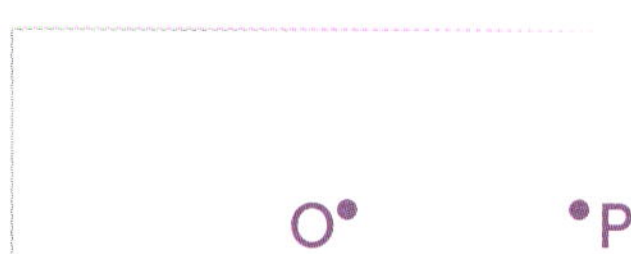

14. 100 × $3.15 = ______

15. 1005 – 8 = ______

Tuesday

Week 22

1. (30 ÷ 5) × (15 ÷ 3) = ______

2. 15:30 $5\frac{1}{2}$ hours later (24-hour time) ☐

3. 4.378 rounded to the nearest hundredth is 4.38, so 2.679 rounded is ______.

4. Write the next four multiples of 4.

148, ______, ______, ______, ______

5. 0.75 = ______ (fraction)

6. How many degrees are in a semicircle?

______

7. Write the missing fractions.

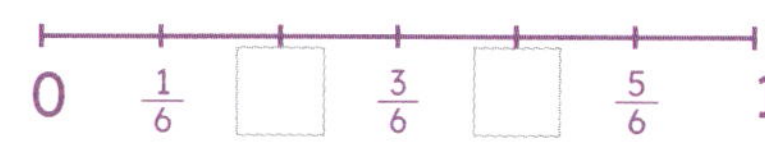

8. 16 ÷ 3 = $\frac{16}{3}$ = ______ (mixed number)

9. 0.52 × 100 = ______

10. 3200 ÷ 80 = ______

11. What is the probability of you receiving homework today? Place a dot on the number line.

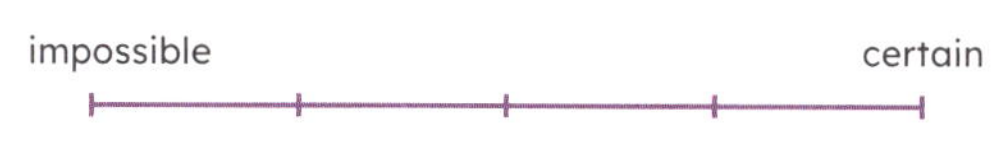

12. 1.5km = ______ m

13. Write in descending order.

$\frac{1}{4}$ $\frac{3}{5}$ $\frac{2}{3}$ $\frac{4}{10}$

______ ______ ______ ______

14. What is the length of A? ______

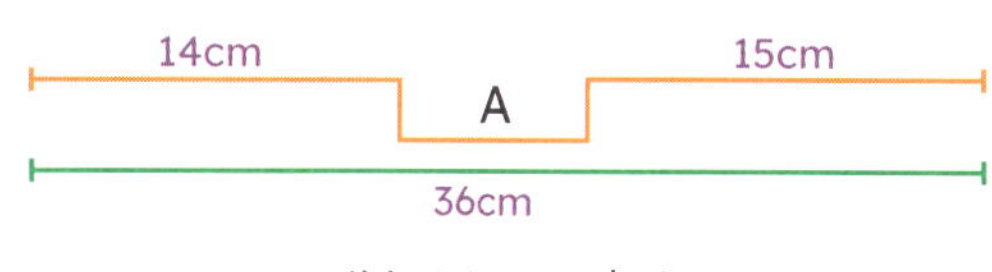

(Not to scale.)

15. There are four rows of eight chairs. How many chairs altogether? (Write as a number sentence.)

______

Week 22

Wednesday

1. (60 ÷ 5) – (4 ÷ 1) = ______

2. If a teacher cuts six whole apples into quarters, and six pieces were eaten, how many apples are left?

   ______ whole apples and ______ quarters.

3. [   ] $4\frac{1}{2}$ hours before [02:30] (24-hour time)

4. Write the ninth of August this year in numerals. ______

5. 1.15, 1.30, 1.45, ______, 1.75

6. The sum of the angles in a square or rectangle is ______.

7. 3500 + 1500 = ______

8. Round $\frac{64}{9}$ to the nearest whole. ______

9. Balance the scales.

   A = 350g

   B = ______kg

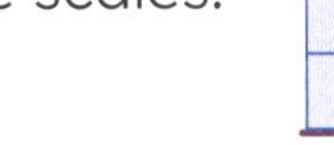

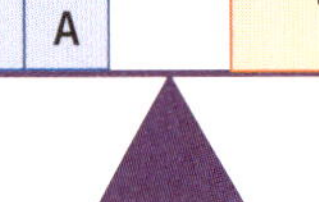

10. Write in ascending order.

    4440 4040 4004 4400

    ______ ______ ______ ______

11. What is the best buy?

    8 USB drives for $48.80 ☐

    6 USB drives for $42 ☐

12. Halve 67. ______

13. A landscaper is designing a garden makeover.

    (a) The maximum width of the paving is ______.

    (b) The area of the paving is ______.

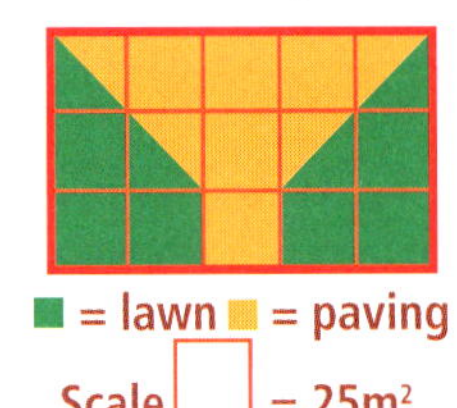

■ = lawn ■ = paving

Scale ☐ = 25m²

14. 0.8 × 3 = ______

15. 10 000 – 50 = ______

Thursday

1. (700 ÷ 7) ÷ (20 ÷ 2) = ______

2. What is the time if the long hand is on 11 and the short hand is between 11 and 12?

   ______

3. Arrange the digits 7, 9, 0, and 4 to make the highest value possible.

   ______

4. 0.29 = ______%

5. Plot these coordinates.

   A = (3,2)

   B = (7,1)

   C = (1,2)

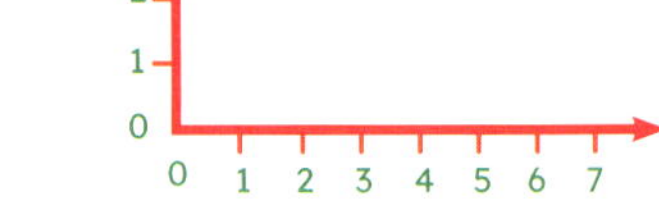

6. 4 × 7 = ______

7. 2 – 0.5 = ______

8. 3005 – 7 = ______

9. 40 + 90 = ______

10. A bus is due to arrive at 3:25 pm. If the time is 2:45 pm, how many minutes will you wait? ______

11. odd + odd + odd = ______

12. Bronte turned over these five cards and mixed them up. What was the probability of randomly selecting a:

    (a) king? ______

    (b) diamond? ______

13. 8 ☐ 7 = 56

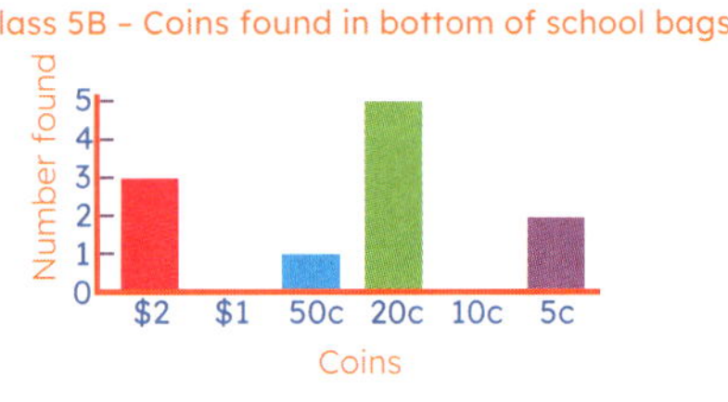

14. Which coins were not found? ______

15. What is the total amount found in the bags? ______

Problem-solving

Ben, Kerry, and Jay bought some sticker packets. Jay bought one more sticker packet than Kerry. Each sticker packet had the same number of stickers in it. In total, Ben bought 40 stickers, Kerry 48 stickers, and Jay 56 stickers.

Let's find out how many stickers were in each packet and how many packets each person bought. (Hint: Start by subtracting Kerry's number of stickers from Jay's.)

Read the question again. **Think** about the information. Underline the important words.

Tick the strategy you will use to work out the answer:

- estimate and check ☐
- look for patterns ☐
- draw a diagram or picture ☐
- construct a table or graph ☐
- use materials ☐
- act it out ☐
- work backwards ☐
- something else. ☐

Solve it:

Reflect on the question and answer.

Check it. Circle another strategy on the list to work out the answer.

Show it:

Friday Review

Week 22

1. 60 + 80 = ______
2. $\frac{35}{100}$ = 0.______ = ______%
3. In Thursday's graph, how many more 20c coins were found than 5c coins? ______
4. a = 5

   9 × a = ______
5. 2000, ______, 12 000, 17 000
6. Write the decimals in ascending order.

   1.19 1.019
   1.009 1.9

   ______ ______ ______ ______
7. If you cut five whole apples into quarters and you eat nine pieces, how many apples are left?

   ______ whole apples and ______ pieces.
8. 2004 – 8 = ______
9. 19 099 > 19 100

   true ☐ false ☐
10. 0.79 × 100 = ______
11. (600 ÷ 10) × (100 ÷ 10) = ______
12. There are three rows of seven school bags. How many bags altogether? (Write as a number sentence.)

    ______
13. 

    $3\frac{1}{2}$ hours before is:

    ☐ (24-hour time)
14. How many degrees are in a straight line? ______
15. 2.5m = ______cm
16. 2400 + 3600 = ______
17. Balance the scales.

    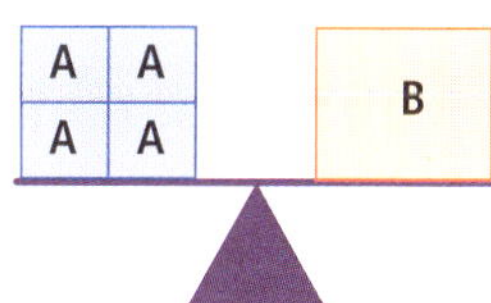

    A = 1.5kg

    B = ______kg
18. The numbers 1 to 20 are in a lotto draw. What is the probability of a multiple of three being chosen? ______

Week 23

Monday

1. What do the letters on this graph represent?

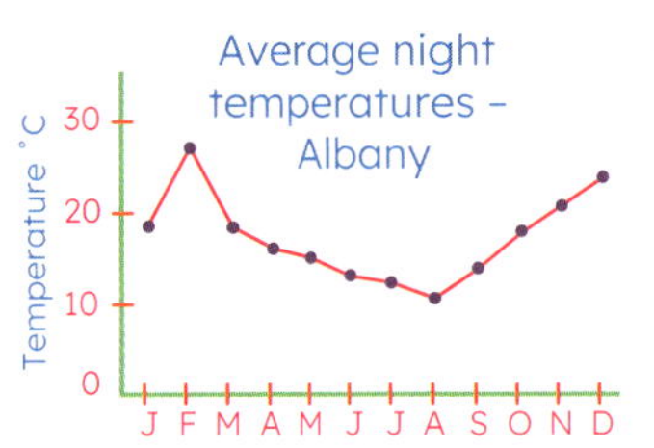

______

2. What is the least amount of coins needed to make $4.15? ______

3. $\frac{4}{8} = \frac{\square}{2}$

4. Draw lines of symmetry for this regular pentagon.

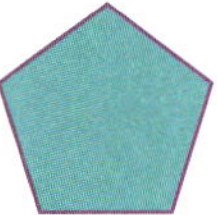

5. 9090 + 10 = ______

6. Round 4.646 to the nearest hundredth.

______

7. $4\overline{)50} = \frac{50}{4} = \frac{25}{2} =$ ______ (mixed number)

8. 410 – 70 = ______

9. Rotate 90° clockwise.

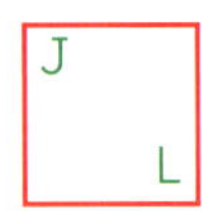

10. Write the previous four multiples of 3.

______, ______, ______, ______, 33

11. Which number is double 38.5?

78 ☐ 77 ☐ 76 ☐ 79 ☐

12. What is the date one week from 24 September?

______

13. What is the missing factor of 36?

1, 2, 3, 4, 6, ______, 12, 18, 36

14. $16.50 + $4.50 = ______

15. In an art sale, a $1200 painting is 20% ($\frac{1}{5}$) off.

Its new price is ______.

Tuesday

1. Using Monday's graph, which month had the warmest nights?

______

2. What four coins can be used to make $3.60?

______, ______, ______, and ______

3. 0.07 + 0.4 + 0.004 = 0.______

4. The XYZ bank charges $5 per month in fees for a savings account. What is the cost per annum (year)?

______

5. The image is translated from coordinates (1,3) to ______.

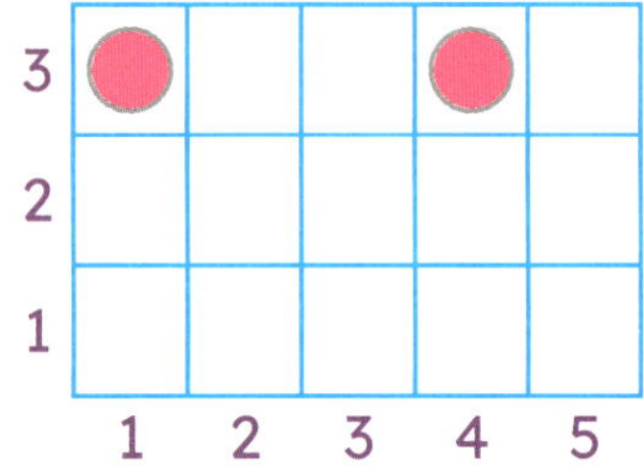

6. $0.5 = \frac{\square}{10}$

7. What is the probability of you flipping a coin and it landing on heads?

______

8. 148mm = ______cm 8mm

9. $\frac{3}{4} > \frac{2}{3}$ true ☐ false ☐

10. 0.008 × 10 = ______

11. The HCF of 15 and 21 is ______.

12. What is the number after 18 999?

______

13. $10\frac{1}{2} - 1\frac{1}{2} =$ ______

14. 250, 500, ______, 1000

15. How many 20g packets will fit in the box?

______

Wednesday

1. Which month had no rainfall? ____

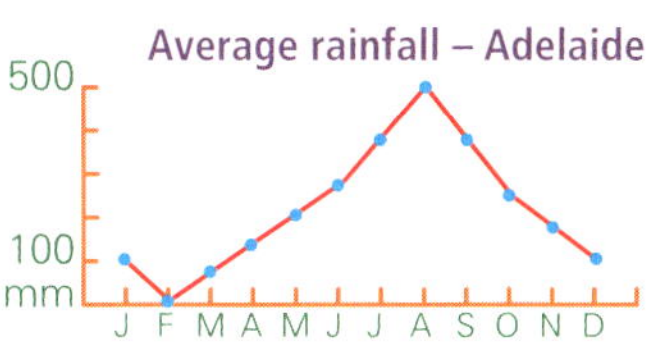

2. 800 000 + 300 000 = ____
3. $31.75 + $9.25 = ____
4. 0.8 × 5 = ____
5. 15, 30, 45, ____, ____, 90
6. Circle the vertex.
7. Write the next four multiples of 4.
   52, ____, ____, ____, ____
8. Round 9.988 to the nearest hundredth. ____
9. 7.0 = 10 × 0.7
   9.0 = 10 × ____
10. 400 × 0.05 = ____
11. Measure the length of $\overline{AB}$. ____mm

12. If your income is $19 367, what range should you tick in this survey?
   - $10 000–$15 000 ☐
   - $15 001–$20 000 ☐
   - $20 001–$25 000 ☐
13. Mia took 8 hours to read 200 pages of their book. What is the average number of pages read each hour? ____
14. What is the number after 19 989? ____
15. Plot and label:
   A at (2,3)
   B at (4,3)
   C at (1,3)

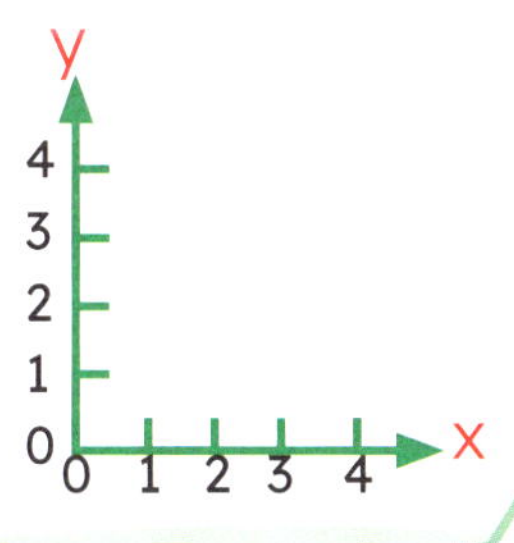

Thursday

1. Using Wednesday's graph, which month was the wettest? ____
2. What is the time 25 minutes after 7:45 am? ____
3. Circle the square number.
   80 81 82 85 90
4. $1.0 = \frac{\square}{10}$
5. How many fifths make up a whole? ____
6. $29.25 + $8.75 = ____
7. $\frac{1}{5}$ of 10 = ____
8. What is the distance from:
   (a) A to D? ____
   (b) A to B? ____
   (c) A to C? ____
   (d) C to D? ____

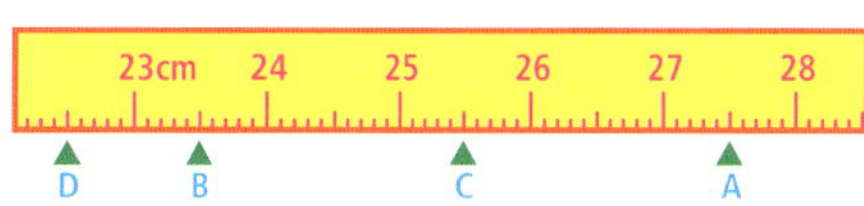

9. 1.5L = ____mL
10. What is the value of the 9 in 4339? ____
11. What is the area of a 4 by 6 grid? ____ squares
12. Is a sphere 2- or 3-dimensional? ____
13. Colour the sides parallel to the red lines.

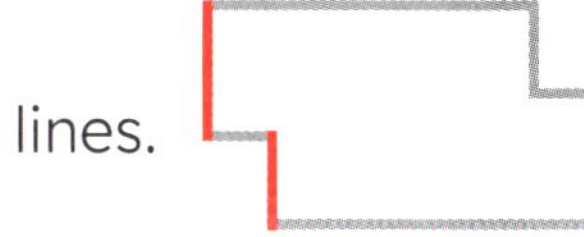

14. $\frac{8}{12} - \frac{4}{12}$ = ____
15. Write four hundred and forty-five thousand, one hundred in numerals. ____

Week 23

Problem-solving

Look at Wednesday's graph that shows annual average rainfall in Adelaide. Choose a different Australian city and research its average annual rainfall. Display the data you found.

Let's find out three differences in annual average rainfall between Adelaide and the city you chose.

Read the question again. **Think** about the information. Underline the important words.

Tick the strategy you will use to work out the answer:

- estimate and check ☐
- look for patterns ☐
- draw a diagram or picture ☐
- construct a table or graph ☐
- use materials ☐
- act it out ☐
- work backwards ☐
- something else. ☐

Solve it:

Reflect on the question and answer.

Check it. Circle another strategy on the list to work out the answer.

Show it:

Friday Review

1. What is the least amount of coins needed to make \$3.80?

   ______

2. \$41.25 + \$8.75

   = ______

3. $0.3 = \frac{\square}{10}$

4. $\frac{1}{4}$ of 200 = ______

5. 204 – 9 = ______

6. 10 × 0.4

   = ______

7. Double 39.5.

   ______

8. 800 × 0.05

   = ______

9. In an art sale, a painting was \$2000. How much discount is 25%?

   ______

10. 77 ÷ 11 = ______

11. $\frac{1}{5} > \frac{1}{2}$

    true ☐ false ☐

12. What is the time 25 minutes after 7:50 pm?

    ______

13. 2900m

    = ______km

14. If you are paid \$90 000 per annum, what would you earn in six months?

    ______

15. Rotate 90° anticlockwise.

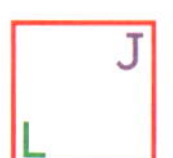

16. 350 – 80

    = ______

17. 129mm

    = ______cm 9mm

18. Which month had the coolest nights?

    ______

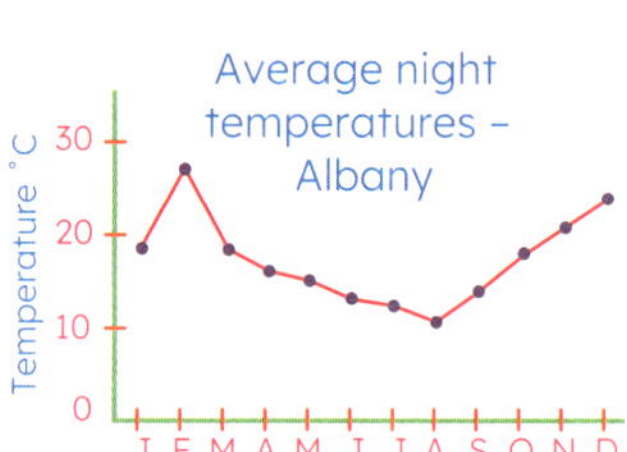

Sp St P

Monday

1. The cross-section made after the cut is what 2D shape?

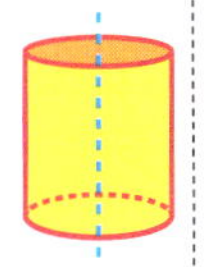

____

2. Colour 350mL.

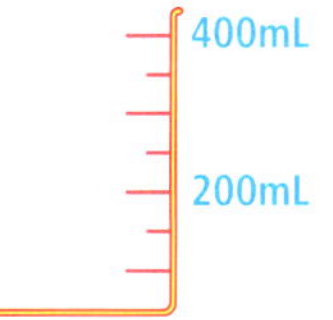

3. 415 – 35 = _______

4. The magic square sums to 15 (using 1 to 9) in all directions. Complete the square.

4		8
	5	
		6

5. If a 1500L water tank is half full, how many litres are remaining?

____

6. 0.4 = $\frac{\square}{10}$

7. Round 2.776 to the nearest hundredth. _______

8. How many angles does an irregular octagon have? _______

9. Complete the multiples of 7.

7	14		28	35
				70

10. If you paid $16.90 for lunch, what change would you have received from $20?

____

11. (a) 5$\overline{)200}$ = _______ (b) 10$\overline{)400}$ = _______

12. $3\frac{1}{2} - 1\frac{1}{2}$ = _______

13. This 3D object is a

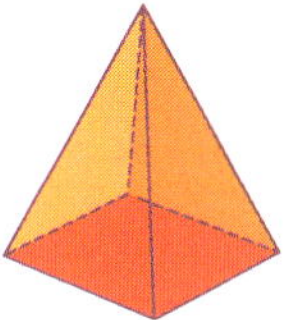

____

_______________.

14. How many vertices does it have? _______

15. What is the time if the long hand is on 10 and the short hand is between 11 and 12?

____

Tuesday

1. The cross-section made after the cut is what 2D shape?

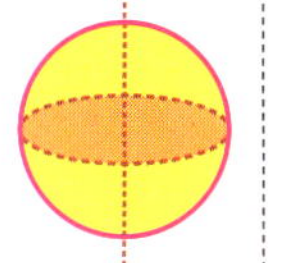

____

2. Which two shapes have translated?

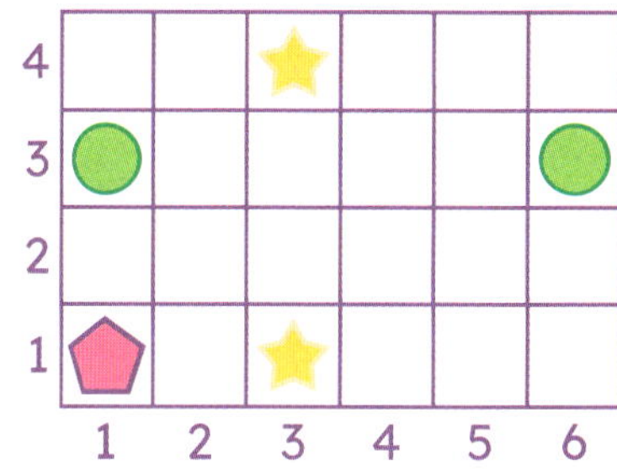

____

____

3. What five coins can be used to make $4.75?

_______, _______, _______, _______, and _______

4. 0.6 + 0.4 = _______

5. If the time is 8:35 am, what will the time be in half an hour?

____

6. 50 × 0.4 = _______

7. Round 27 591 to the nearest ten thousand. _______

8. Match the times (non-daylight savings).

Warrane / Sydney •	• 4:45 pm
Tarndanya / Adelaide •	• 2:45 pm
Boorloo / Perth •	• 4:45 pm
Naarm / Melbourne •	• 4:15 pm

9. 9.5, _______, 10.5, 11, 11.5

10. Draw 6 lines of symmetry on this regular hexagon.

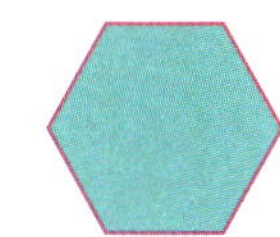

11. In a hat are 4 red pens, 2 blue pens, and 2 green pens. What is the probability of picking a green pen?

____

12. The HCF of 12 and 15 is _______.

13. _______ + 15 = 90

14. 410 - 40 = _______

15. Write nine hundred and fifty thousand in numerals.

____

Week 24

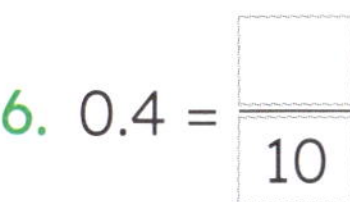

Week 24

Wednesday

1. The cross-section made after the cut is what 2D shape?

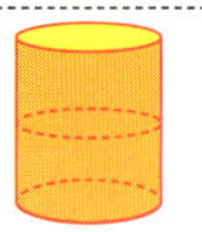

____

2. $50 \times 0.6 =$ ______
3. Share $120.00 equally among 10 people. ______________
4. $n + 1 = 7$, so $n =$ ______.
5. $14 \div 8 =$ ______ r ______
6. $430 - 50 =$ __________
7. 176mm = ________cm 6mm
8. 9 ☐ 3 = 3
9. Tradie Tom measured lengths of pipe from zero to the five points marked. Record the measurements in mm.

   A = ________mm   B = ________mm

   C = ________mm   D = ________mm

   E = ________mm

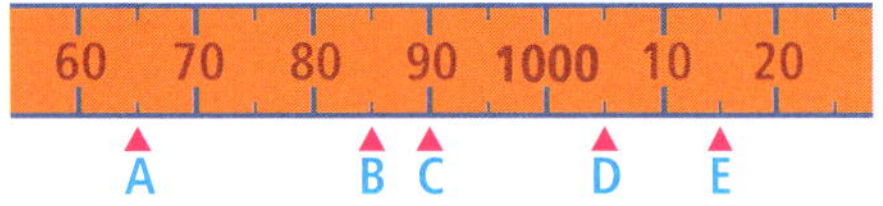

10. $1.75 + $1.25 = __________
11. A rectangle has rotational symmetry to the order of

   ______.

12. Mila read 320 pages of a book in eight hours. What is the average number of pages read per hour?

   ____

13. $805 \div 10 =$ ______
14. In which set are the numbers multiples of 4 and 6?

   12, 24, 28 ☐ 12, 24, 32 ☐ 12, 24, 36 ☐

15. $\frac{6}{10} = \frac{\square}{20}$

Thursday

1. If a cube has a corner cut away, what 2D shape is the cross-section?

____

2. Record the time shown as 24-hour time.

   ____

3. $39 \times 50 \times 2 =$ __________
4. Draw and label:

   (a) a vertical line (V).

   (b) a horizontal line (H).

5. (a) $999 + 8 =$ __________

   (b) $9999 + 8 =$ __________

6. Double 350. ________
7. What is the value of 2 in 6.002? ________
8. $\frac{3}{4} < 0.8$ true ☐ false ☐
9. (a) $\frac{9}{1000} = 0.$________ (b) $\frac{23}{1000} = 0.$________
10. What is the size of x?

   ____

11. $\frac{1}{4}$ of 1000 = ________
12. How many people liked tomatoes?

   ____

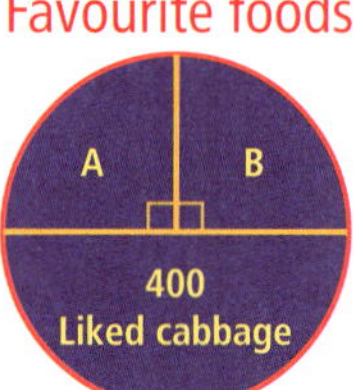

13. 12 ☐ 3 = 9
14. What is the cost of 1kg of butter at $1.50 per 250g? __________
15. $25 \times 12 =$

   $25 \times 4 \times 3 =$ __________

Problem-solving

Daku is cutting cross-sections of 3D objects to find which ones make a rectangle.

Let's draw the nets of five 3D objects that give a rectangular cross-section that Daku may have found.

Read the question again. **Think** about the information. Underline the important words.

Tick the strategy you will use to work out the answer:

- estimate and check ☐
- look for patterns ☐
- draw a diagram or picture ☐
- construct a table or graph ☐
- use materials ☐
- act it out ☐
- work backwards ☐
- something else. ☐

Solve it:

Reflect on the question and answer.

Check it. Circle another strategy on the list to work out the answer.

Show it:

Friday Review

Week 24

1. 997 + 8 = ______
   9997 + 8
   = ______
2. 402 – 9 = ______
3. Share \$150.00 equally among 10 people.
   \$______ each
4. 70, 63, 56, 49,
   ______
5. 8 ☐ 9 = 72
6. $\frac{3}{1000}$ = 0.______
7. $n \div 4 = 3$,
   so n = ______.
8. $5\overline{)300}$ = ______
9. $5\frac{1}{2} - 2\frac{1}{2}$ = ______
10. 10 × \$1.35
    = ______
11. 23:25
    Half an hour later is:
    ☐ (24-hour time)
12. 195mm
    = ______cm 5mm
13. What is the measurement at:
    A? ______mm
    B? ______mm

    80 90 300 10 20
    B A
14. This is an irregular
    ______.

15. Colour 75mL.

    100mL

16. The cross-section made after the cut is what 2D shape?
    ______

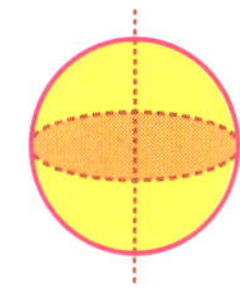

17. Look back at the pie graph from Thursday. What is the percentage that liked potatoes?
    ______
18. A bag of marbles contains six purple, three green, and three blue marbles.

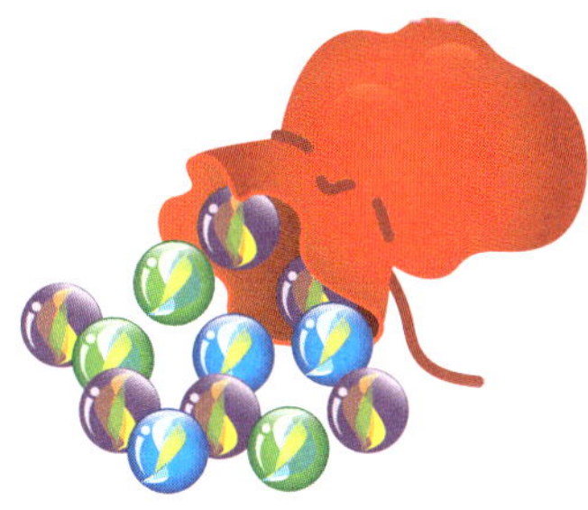

What is the probability of choosing a green marble?

$\frac{3}{12}$ ☐ $\frac{6}{12}$ ☐

$\frac{1}{10}$ ☐ $\frac{9}{12}$ ☐

N     P 

Week 25

Monday

1. 0.5 > 1 true ☐ false ☐
2. 9997 + 9 = ______
3. 25 × 16 = 25 × 4 × 4 =

   ______ × 4 = ______
4. Write $\frac{1}{4}$ as a decimal. ______
5. Share $210.00 equally among 21 people.

   ______
6. What is the product of 3 and 6? ______
7. $4.25 + $7.75 = ______
8. Add 74 tens and 35 ones. ______
9. 652 ÷ 4

   600 ÷ 4 = ______

   40 ÷ 4 = ______

   12 ÷ 4 = ______

   Total = ______
10. 10 000 + 39 = ______
11. How many minutes' difference? ______

12. Round 7.334 to the nearest hundredth.

    ______
13. Neda had a 6 × 6 square in tiles. How many tiles are missing from the square?

    ______

14. How many weeks are in a year? ______
15. What is the place value of 2 in 27 500?

    ______

Tuesday

1. 3789 > 3699 true ☐ false ☐
2. 400 – 125 = ______
3. Write $\frac{3}{4}$ as a decimal. ______
4. 3.1 – 0.8 = ______
5. Draw the top view.

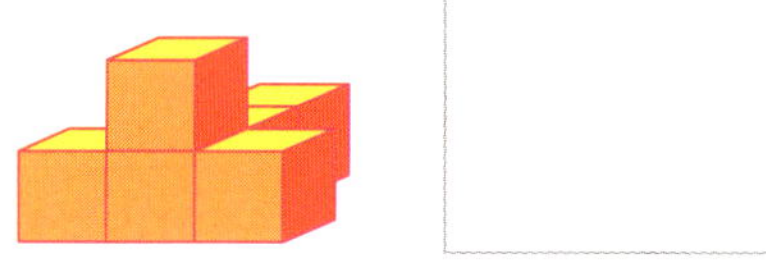

6. a = 6

   a × 7 = ______
7. 400 ÷ 4 = ______
8. 60 × 8 = 480

   59 × 8 = 472

   58 × 8 = ______
9. odd × odd = ______
10. What is the length of this pencil?

    ______ mm

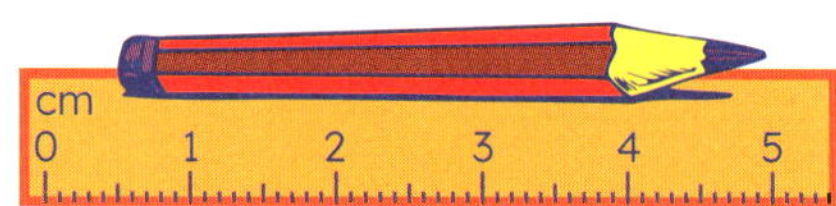

11. In which set are the numbers multiples of 3 and 4?

    9, 12, 16 ☐ 8, 12, 16 ☐

    12, 24, 36 ☐ 18, 24, 36 ☐
12. 44 × 7 = (______ × 7) + (______ × 7)

    = ______ + ______

    = ______
13. 3.7, 3.8, 3.9, ______
14. 2173 – 74 = ______
15. From (1,2) label all the grid squares north-east as Z.

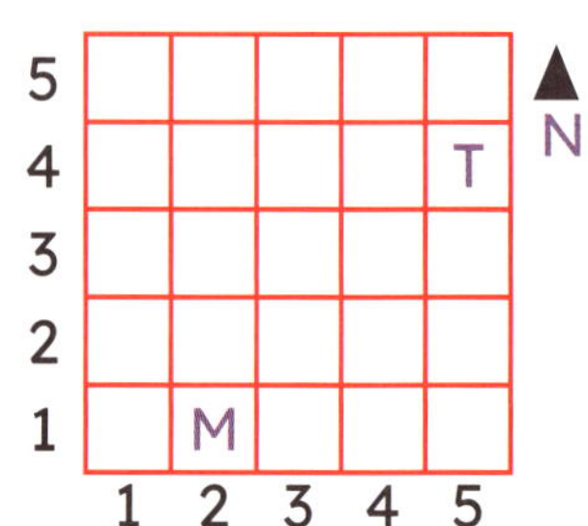

Wednesday

1. $\frac{1}{2} < 1.3$ true ☐ false ☐

2. 500 – 125 = ______

3. $\frac{1}{5}$ of 420 = ______

4. What is the date one week before 5 April?

______

5. The cross-section made after the cut is what 2D shape?

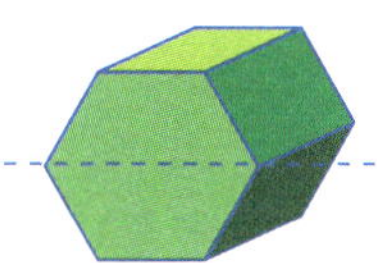

______

6. $\frac{9}{10} + \frac{5}{10} = \frac{\square}{10}$ = ______ (decimal)

7. How many degrees in a right angle?

______

8. 1000 + 23 = ______

9. \$8.75 + \$13.25 = ______

10. 2.4km = ______ km ______ m

11. Jess recorded the sum of a pair of dice for 10 throws: 4, 7, 9, 7, 11, 10, 8, 7, 6, 10.

(a) Order the numbers from smallest to largest.

______

(b) Find the mode.

______

12. What are the missing factors of 24?

1, 2, 3, ______, ______, 8, 12, 24

13. 25 000 + 75 000 = ______

14. Emily received \$12.50 pocket money each week.

After 20 payments Emily had earned ______.

15. What is the perimeter of a regular hexagon with 10cm sides?

______

Thursday

Week 25

1. 0.002 = $\frac{2}{10}$ ☐ $\frac{2}{100}$ ☐ $\frac{2}{1000}$ ☐

2. How many minutes' difference?

______

3. 10 000 – 34 = ______

4. $5\overline{)260}$ = ______

5. How many scored less than 20 runs?

______

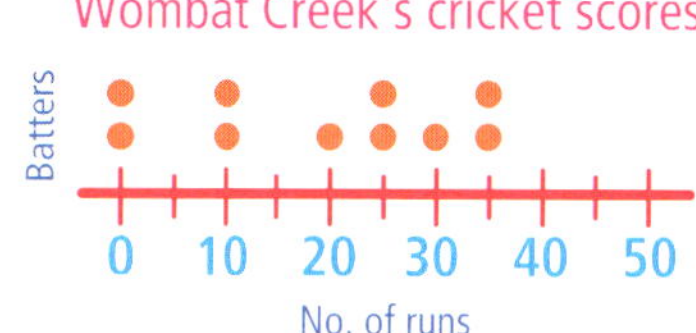

6. even + even + odd = ______

7. 2.2 – 0.8 = ______

8. 199 997, 199 998, 199 999, ______

9. How many people preferred Sorrento and Narrows Bridge?

______

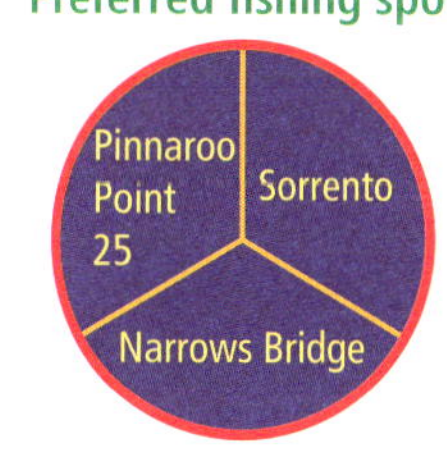

10. 0.4 + 0.9 = ______

11. What is the area of a 7 by 4 grid?

______ squares

12. Write one hundred and ten thousand and ten in numerals.

______

13. 710 ÷ 10 = ______

14. Write A at grid coordinates (1,3).

15. The direction from A to B is:

south-west. ☐

south-east. ☐

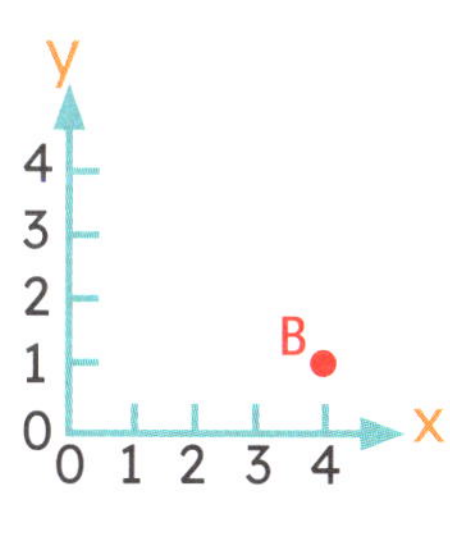

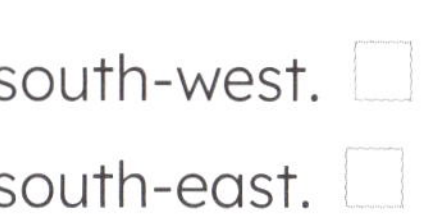

Week 25

Problem-solving

Carla rescued these four injured animals and weighed them.

Animal	Weight
koala	15.25kg
wombat	20 500g
quokka	4.35kg
possum	1200g

Using the symbols < and/or >, let's help Carla write the weights of the rescue animals in order from lightest to heaviest.

Read the question again. **Think** about the information. Underline the important words.

Tick the strategy you will use to work out the answer:

- estimate and check ☐
- look for patterns ☐
- draw a diagram or picture ☐
- construct a table or graph ☐
- use materials ☐
- act it out ☐
- work backwards ☐
- something else. ☐

Solve it:

Reflect on the question and answer.

Check it. Circle another strategy on the list to work out the answer.

Show it:

Friday Review

1. $\frac{1}{2} > 0.45$

   true ☐ false ☐

2. 150 – 70 = ______

3. What is the probability of an odd number?

   ______

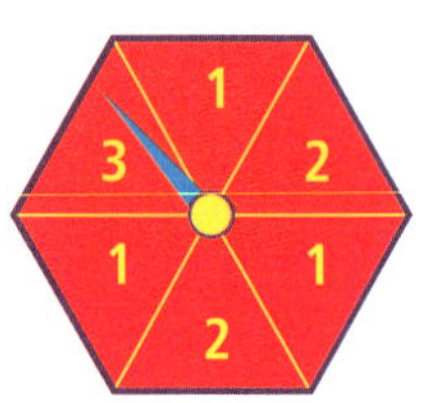

4. $\frac{6}{10} + \frac{3}{10} = \frac{\square}{10}$

   = ______ (decimal)

5. Add 38 tens and 25 ones.

   ______

6. 25, 75, 125, 175, ______

7. 0.008 =

   $\frac{8}{100}$ ☐ $\frac{8}{1000}$ ☐

   $\frac{8}{10}$ ☐ $\frac{8}{1}$ ☐

8. What is the product of 6 and 8?

   ______

9. Round 3.936 to the nearest hundredth.

   ______

10. 75% = 0.______

11. Halve 90. ______

12. odd × even

    = ______

13. How many minutes' difference?

    ______

    1:35 → 8:15

14. 2.7km =

    ______km ______m

15. What 2D shape will you see in the cross-section?

    ______

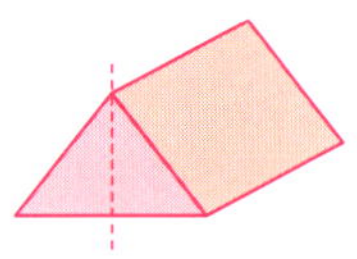

16. 3096 – 97

    = ______

17. Look back at Tuesday's grid. In which direction are you travelling if you go from (2,1) M to (5,4) T?

    ______

18. What is the total number of popped balloons?

    ______

Balloons popped at Poppy's party

Balloons popped

Pink Red Blue Green

N A M Sp St P

Monday

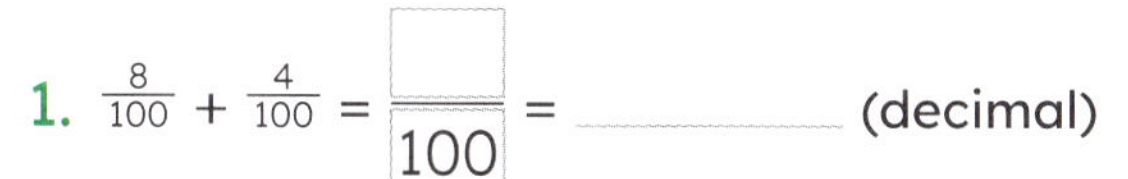

1. $\frac{8}{100} + \frac{4}{100} = \frac{\square}{100}$ = ______ (decimal)

2. A landscaper marked an area of 100m × 100m = ______ m^2 or ______ ha.

(Not to scale.)

3. The time is 7:30 pm. What will the time be in $5\frac{1}{2}$ hours? ______

4. 90 ÷ 10 = ______

5. 90 000 + 110 000 = ______

6. odd + even + odd = ______

7. Halve 300. ______

8. $\frac{1}{2} = \frac{\square}{4}$

9. 0.6 × 6 = ______

10. Which angle is closest to 30°? ______

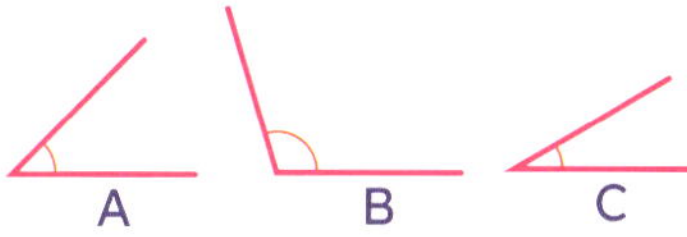

11. 48 ☐ 8 = 6

12. Chef found that one cup of flour = 100g. What is the mass of $3\frac{1}{4}$ cups of flour? ______

13. Match the times (non-daylight savings).

Tarndanya / Adelaide	• —— •	5:00 pm
Boorloo / Perth	• •	5:30 pm
Warrane / Sydney	• •	3:30 pm
Gulumoerrgin / Darwin	• •	5:00 pm

14. $\frac{1}{4}$ of 900 = ______

15. How many lines of symmetry does the triangle have? ______

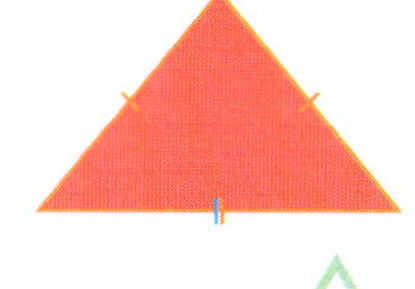

Tuesday

Week 26

1. $\frac{8}{10} + \frac{7}{10}$ = ______ $= \frac{\square}{10}$ = ______ (decimal)

2. Add the next five throws (results) to the dot plot graph:

(4, 3), (2, 5), (2, 1), (6, 3), (1, 3).

Results of throwing 2 dice

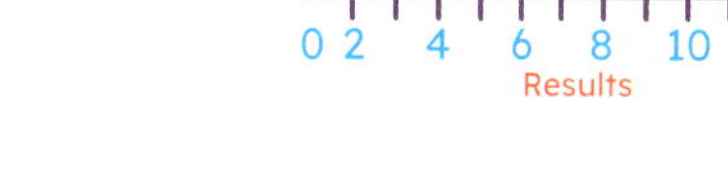

3. 0.8 + 0.5 = ______

4. Round 7.296 to the nearest hundredth. ______

5. 505 – 8 = ______

6. \$20.00 – \$10.50 = ______

7. This is a ______.

8. Which equation is equal to 12 × 7?

8.6 × 10 = 86	☐	7.2 × 10 = 72	☐
8.4 × 10 = 84	☐	8.2 × 10 = 82	☐

9. $4\frac{1}{2} + 9\frac{1}{2}$ = ______

10. How many fifths make up 3 wholes? ______

11. If you have ridden 17km of a 40km bicycle race, how far do you have left to go? ______

12. Colour the sides parallel to the red line.

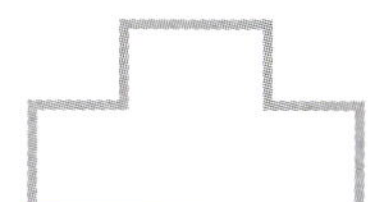

13. \$1.95 × 100 = ______

14. 20 × 7 = 140

19 × 7 = ______

18 × 7 = ______

15. 572 596 – ______ = 500 000

Wednesday

1. $\frac{10}{100} + \frac{90}{100} = \frac{\square}{100}$ = ________ (decimal)

2. Packets of jelly beans are in 500g boxes. What is the total number of packets in boxes A and B?

____

3. What is the HCF of 18 and 27? ________

4. 2.4 + 0.6 = ________

5. Lara took 24 selfies each day. A total of 192 selfies were taken over ________ days.

6. This is a net of a ________.

7. 1 100 000 – 400 000 = ________

8. 139mm = ________cm 9mm

9. 60 000 + 80 000 = ________

10. 105, 90, 75, ________, 45, 30

11. Which measurement would you use to measure milk for a cake recipe?

mL ☐ kg ☐ cm ☐ m ☐

12. Match each sum with its decimal.

10 × 0.1 •	• 0.01
10 × 0.01 •	• 1.0
10 × 0.001 •	• 0.1

13. 700 + 8000 + 60 + 2 = ________

14. Luke, a farmer, sold 1t of grapes for $1000. If the grapes were packed in 25kg boxes, how many boxes did Luke sell? (1t = 1000kg)

____

15. How much did each box of grapes sell for? ________

Thursday

1. $\frac{4}{10} + \frac{4}{10} = \frac{\square}{10}$ = ________ (decimal)

2. a = 7

a + 9 = ________

3. $\frac{1}{5} > \frac{1}{3}$ true ☐ false ☐

4. Draw the top view.

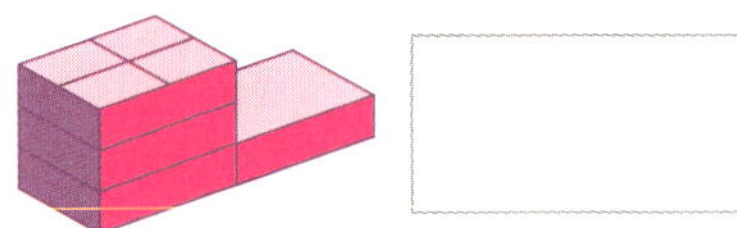

5. 25 × 12 × 4 = ________

6. Halve 700. ________

7. ________, 9.9, 9.8, 9.7

8. Match each object with its number of edges.

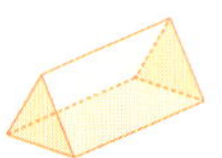 • • 8

 • • 12

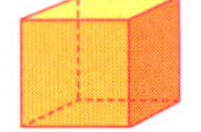 • • 9

9. $\frac{1}{4}$ of 40 = ________

10. 36 ÷ 6 = ________ ÷ 3

11. Which set has two prime numbers?

3, 9 ☐ 2, 8 ☐ 2, 7 ☐ 1, 5 ☐

12. 4 × 8 = 8 + 8 + 8 + 8 = ________

13. The number on the abacus is ________.

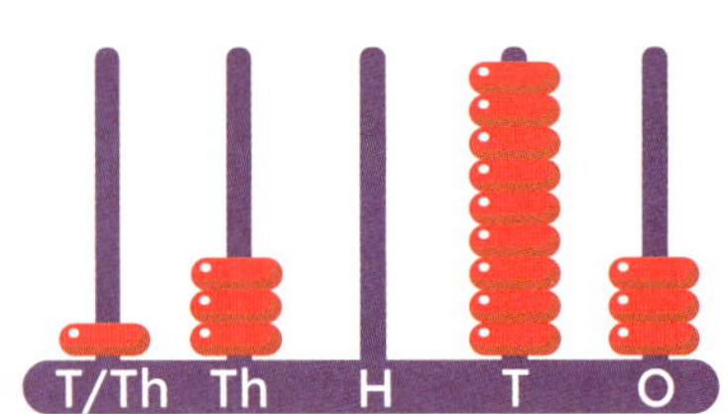

14. 1.6L = ________mL

15. 40 × 8 = 320 39 × 8 = ________

38 × 8 = ________ 37 × 8 = ________

Problem-solving

There were 100 chickpeas in a jar. Alice chomped on 20, Allira gobbled down 15, and Arika scoffed half the number that Alice ate.

Let's find out how many chickpeas they ate in total, as a fraction and a decimal.

Read the question again. **Think** about the information. Underline the important words.

Tick the strategy you will use to work out the answer:

- estimate and check ☐
- look for patterns ☐
- draw a diagram or picture ☐
- construct a table or graph ☐
- use materials ☐
- act it out ☐
- work backwards ☐
- something else. ☐

Solve it:

Reflect on the question and answer.

Check it. Circle another strategy on the list to work out the answer.

Show it:

Friday Review

Week 26

1. 24 ÷ 3 = ______ ÷ 6
2. $b - a = 5$
   $a = 6$
   $b =$ ______
3. 405 – 9 = ______
4. 2.2 + 0.8 = ______
5. 18 099 > 18 101
   true ☐ false ☐
6. $4\overline{)520}$ = ______
7. 1 200 000 – 800 000
   = ______
8. $\frac{6}{100} + \frac{7}{100} = \frac{\square}{100}$
   = ______ (decimal)
9. 25 × 18 × 4
   = ______
10. Write forty-two thousand and one hundred in numerals.
    ______
11. $2.55 × 100
    = ______
12. Amy took 18 selfies each day. A total of 144 selfies were taken over ______ days.
13. 19 997, 19 998, 19 999, ______
14. 1ha = ______ m^2
15. Colour:
    (a) 16mL in pink.
    (b) 4mL more in blue.

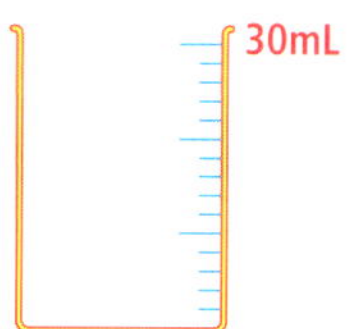

16. What is the total number of packets in A and B?

______ packets

17. The time in Warrane / Sydney is noon. What is the time in Western Australia?
    ______
18. Look at Tuesday's graph again. What is the mode of the data?
    ______

P

Monday

1. Match each mixed number with its improper fraction.

$4\frac{4}{5}$	$\frac{26}{5}$
$3\frac{3}{5}$	$\frac{24}{5}$
$5\frac{1}{5}$	$\frac{18}{5}$

2. 180 students were asked, 'What is a great family pet?' How many chose cats? ______

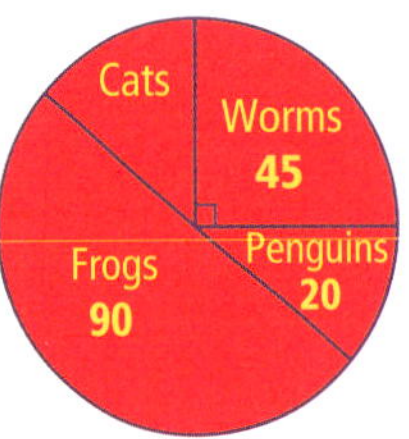

3. 16 ÷ 7 = ______ r ______

4. 396 295 – ______ = 390 000

5. How many days are in a year? ______

6. 31 + 29 = ______

7. What is the value of the 4 in 470 000? ______

8. Draw a reflection of:

9. Round 17 500 to the nearest ten thousand. ______

10. $\frac{1}{2} > \frac{2}{3}$ true ☐ false ☐

11. 18, 25, 32, 39, ______, 53

12. Draw the top view.

13. Circle the multiples of 6.

22 24 30 32 44 42 54 56

14. Write forty-nine thousand, four hundred and ninety in numerals.

______

15. A fast train travels at 380km/h. How far would it travel in half an hour?

______

Tuesday

1. Match each improper fraction with its mixed number.

$\frac{8}{3}$	$1\frac{1}{3}$
$\frac{4}{3}$	$3\frac{2}{3}$
$\frac{11}{3}$	$2\frac{2}{3}$

2. 6 × 7 = 42 42 ÷ 7 = ______

3. 230 + 80 = ______

4. 0.8 × 4 = ______

5. $\frac{1}{4}$ = ______ %

6. The factors of 18 are:

1, ______, ______, ______, ______, 18

7. Join the dots to make a cube.

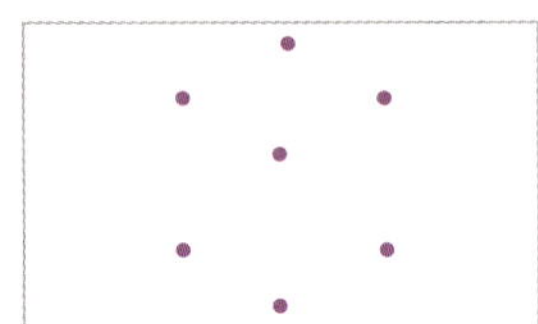

8. Double 11 500. ______

9. Round 6.7 to the nearest whole number. ______

10. 25 × 28 = ?

25 × 4 = ______

100 × ______ = ______

11. Halve 11 000. ______

12. 0.07 + 0.9 + 0.004 = ______

13. The hexagon has a rotational symmetry of:

2 ☐ 4 ☐ 6 ☐

14. even + odd + even = ______

15. A bag of marbles contains three red marbles, four blue marbles, two green marbles, and one yellow marble.

What is the probability of choosing a blue?

$\frac{4}{10}$ ☐ $\frac{3}{10}$ ☐ $\frac{5}{10}$ ☐ $\frac{10}{10}$ ☐

Wednesday

1. Match each improper fraction with its mixed number.

$\frac{31}{5}$	$4\frac{4}{5}$
$\frac{24}{5}$	$5\frac{3}{5}$
$\frac{28}{5}$	$6\frac{1}{5}$

2. Draw an irregular pentagon.

3. (a) 17 – 9 = ______ (b) 1.7 – 0.9 = ______
4. $10 – $3.90 = ______
5. 73 + 38 = ______
6. $\frac{9}{10} + \frac{8}{10} = \frac{\square}{10}$ = ______ (decimal)
7. 43 000 + 9000 = ______
8. 7089 – 99 = ______
9. How many laps of a 50-metre pool would have to be swum to complete 2.5km?

   ______ laps

10. 6⟌804 = ______
11. The painting was $1500. How much discount is 10%?

   ______.

12. (40 ÷ 4) × (60 ÷ 5) = ______
13. If the average car length is 4m, what is the approximate length of a single-lane traffic jam with 250 cars? (Assume bumper to bumper.)

   = ______

14. 6.0 ÷ 5 = ______ ÷ 10
15. 1001 – 7 = ______

   10 001 – 7 = ______

Thursday

Week 27

1. Match each improper fraction with its mixed number.

$\frac{7}{4}$	$2\frac{1}{5}$
$\frac{11}{5}$	$2\frac{3}{4}$
$\frac{11}{4}$	$1\frac{3}{4}$

2. 25 × 28 × 4 = ______
3. 9 × 0.3 = ______
4. Ivan read 120 pages of a book in two hours. How many pages were read each minute?

   ______

5. $20 – $9.50 = ______
6. 8 × 0.4 = 16 × ______ = ______
7. A grocery bag contains 10 potatoes, 6 bananas, 8 apples, and 6 pears. What fraction of the groceries is potatoes?

   ______

8. 8 × 9 = ______
9. 2600 + 800 = ______
10. What is the time difference? ______

11. Halve $\frac{1}{2}$. ______
12. The HCF of 14 and 21 is ______.
13. 500, 2000, 3500, ______, 6500
14. Zoe recorded 12 spins: 2, 4, 2, 3, 1, 1, 3, 4, 1, 2, 3, 2.

   (a) Order the numbers from smallest to largest.

   ______

   (b) Find the mode.

   ______

15. Draw beads on the abacus to show 12 054.

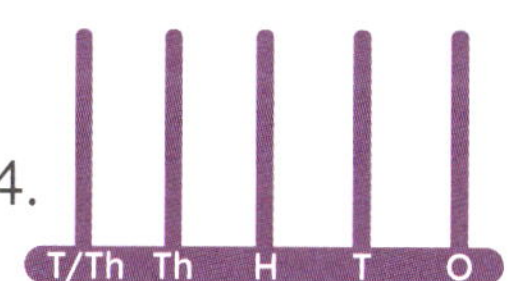

Week 27

Problem-solving

Five students were having a pie eating competition. Ben ate $2\frac{1}{4}$ pies, Den ate $\frac{5}{4}$, Hen ate $\frac{13}{4}$, Jen ate $3\frac{3}{4}$, and Ken ate $\frac{9}{4}$.

Let's find out which two students ate the same amount of pie and how many whole pies were eaten altogether.

Read the question again. **Think** about the information. Underline the important words.

Tick the strategy you will use to work out the answer:

- estimate and check ☐
- look for patterns ☐
- draw a diagram or picture ☐
- construct a table or graph ☐
- use materials ☐
- act it out ☐
- work backwards ☐
- something else. ☐

Solve it:

Reflect on the question and answer.

Check it. Circle another strategy on the list to work out the answer.

Show it:

Friday Review

1. Match each mixed number with its improper fraction.

$4\frac{1}{3}$ 

$5\frac{2}{3}$ 

$2\frac{1}{3}$ 

2. 2003 – 7

= ________

3. Halve 16 750.

____

4. Draw the beads to show 10 101.

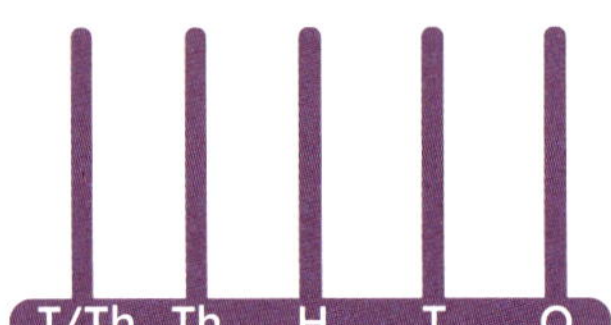

5. The painting was $280, now it costs

$________.

6. 55 000 + 9000

= ________

7. $20 – $7.50

= ________

8. What is the HCF of 18 and 24?

____

9. $6\overline{)534}$ = ________

10. 100 × 8 = 800

99 × 8 = 792

98 × 8 = ________

11. 9 × 0.5 = ________

12. Double 59.

____

13. Write the eighteenth of September as a numeral.

____

14. This is an irregular ________.

15. Plot A at coordinate (1,4).

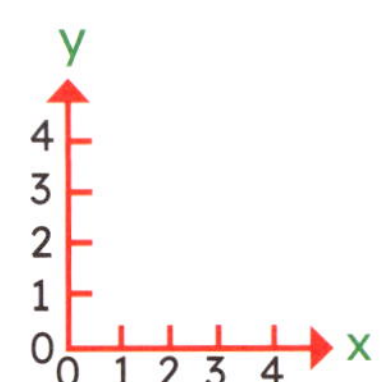

16. In Monday's pie graph, how many more preferred frogs than worms?

____

17. A super-fast train travels at 360km/h. How far does it travel in half an hour?

____

18. In Tuesday's marble bag (question 15), what is the probability of NOT choosing a yellow?

________ in ________

Monday

1. 17.6 = 10 + 7 + 0.6

   21.3 = ______ + ______ + ______

2. Using a ruler, reduce this line by half and write the new measurement.

   ______

3. 2090 = ______ tens

4. How many days make up March and April?

   ______

5. Round 15 960 to the nearest thousand.

   ______

6. 400 – 11 = ______

7. Order $\frac{1}{4}$, $\frac{1}{8}$, $\frac{1}{5}$,

   0, ______, ______, ______, 1

8. Share $20.00 equally among 8 people.

   ______

9. (a) $1 - \frac{1}{2}$ = ______ (b) $2 - \frac{1}{3}$ = ______

10. 79 + 44 = ______

11. What is the time if the long hand is on 9 and the short hand is between 2 and 3?

    ______

12. 0.92 + 0.10 = ______

13. 6 × 300 = ______

14. Jasmine has a new book and reads, on average, 14 pages every 20 minutes for two hours. How many pages have been read?

    ______

15. 600 + 900 = ______

Tuesday

Week 28

1. 32.45 = 30 + 2 + 0.4 + 0.05

   28.37 = ______ + ______ + ______ + ______

2. A triangular pyramid has

   ______ faces and ______ edges.

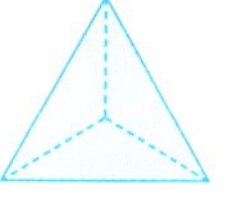

3. 9 × 4 = ______

4. $20.00 – $14.40 = ______

5. 1.1 × 5 = ______

6. 9370 = ______ hundreds ______ tens

7. What is the interval between bus departures?

   ______

Departures
9:03 am
9:28 am
9:53 am

8. Rotate 180°.

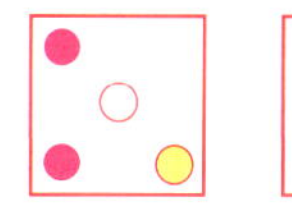

9. 800 – 50 = ______

10. 425 000 + 9000 = ______

11. 27 ÷ 5 = ______ r ______ or ______ (decimal)

12. (9 × 5) ÷ (9 × 1) = ______

13. Using a ruler, reduce this line by half and write the new measurement.

    ______

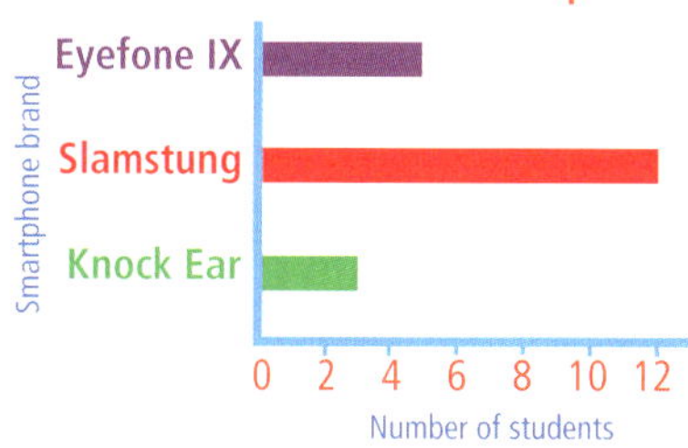

What fraction of students liked:

14. Slamstung? ______

15. Knock Ear? ______

Week 28

Wednesday

1. $70 + 3 + 0.2 + 0.05 =$ ______

2. Rehka needs 120mL of lemon juice for a lemon meringue pie. Colour that amount.

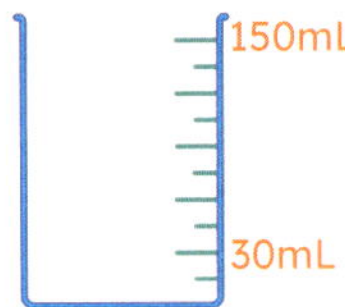

3. $4 \times 8 =$ ______ $8 \times 8 =$ ______

4. $9000 + 299\,000 =$ ______

5. $81 \div 9 =$ ______

6. Share $100.00 equally among five people.

______

7. Write in ascending order.

$\frac{1}{2}$ $\frac{1}{5}$ $\frac{1}{8}$ $\frac{1}{3}$

______ ______ ______ ______

8. Rotate 270° clockwise.

9. The factors of 21 are:

1, ______, ______, 21

10. Round 6.725 to the nearest tenth. ______

11. $(65 + 15) - (40 \div 8) =$ ______

12. If you cut 3 pizzas into sixths, how many pieces will you have? ______

13. $2 - \frac{3}{4} =$ ______

14. Write $\frac{1}{4}$ as a decimal. ______

15. Packets of jelly beans are packed in 500g boxes. What is the total number of packets in boxes A, B, C, and D?

______

Thursday

1. $40 + 2 + 0.9 + 0.06 =$ ______

2. What is the interval between bus departures?

Departures
7:06 am
7:26 am
7:46 am
8:06 am

______

3. How many days make up July and August? ______

4. $9 \div 2 = 4.5$ $11 \div 2 =$ ______

5. $10\,005 - 105 =$ ______

6. Draw ↓ to show the heavier side.

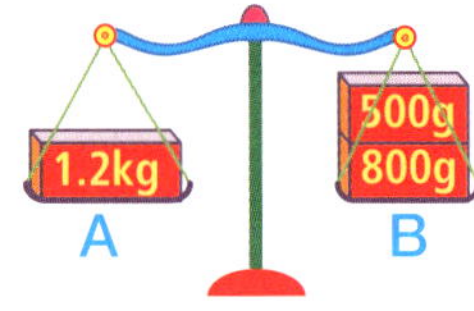

7. $3 - \frac{2}{5} =$ ______

8. $\frac{3}{4} =$ ______%

9. $402 \times 3 =$ ______

10. If you buy two hot chocolates at $1.80 each, what change do you receive from $5.00?

______

11. Plot the coordinates.

(4,3) – A

(3,4) – B

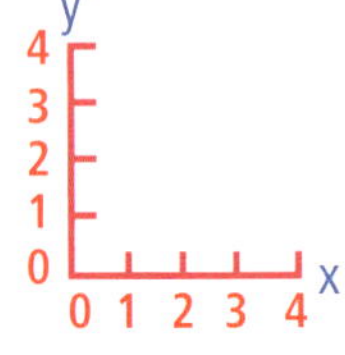

12. The factors of 30 are:

1, ______, ______, ______, ______, ______, 15, 30

13. The results of the last 12 spins are:

2, 3, 7, 7, 13, 7, 11, 13, 13, 2, 11, 5

The chance of spinning a 5 next is:

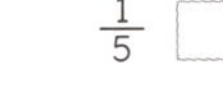

1 in 6 or $\frac{1}{6}$ ☐ $\frac{1}{5}$ ☐

highly likely to certain ☐

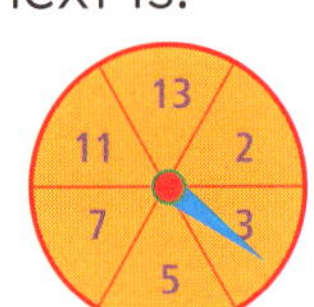

14. If you had swum 2.1km in a 50m pool, how many laps would that be? ______

15. Halve 51. ______

Problem-solving

Four students had a ball throwing competition. Pam threw the ball 20 + 5 + 0.9 + 0.07 metres, Gem threw 4 + 0.06 + 20 + 0.2 metres, Kim threw 5 + 0.4 + 0.01 + 20 metres, and Jim threw 0.2 + 4 + 20 + 0.09 metres.

Let's find out which student threw the ball the furthest and how far they threw it.

Read the question again. **Think** about the information. Underline the important words.

Tick the strategy you will use to work out the answer:

- estimate and check ☐
- look for patterns ☐
- draw a diagram or picture ☐
- construct a table or graph ☐
- use materials ☐
- act it out ☐
- work backwards ☐
- something else. ☐

Solve it:

Reflect on the question and answer.

Check it. Circle another strategy on the list to work out the answer.

Show it:

Friday Review

Week 28

1. 13 ÷ 2 = ______ r ______ or ______ (decimal)
2. 104, 96, 88, 80, ______, ______, ______
3. 755 000 + 8000 = ______
4. $5\overline{)105}$ = ______
5. 8001 – 101 = ______
6. 24.56 = ______ + ______ + 0.______ + 0.______
7. In Tuesday's graph, how many students were surveyed? ______
8. 4 × 700 = ______
9. $3 - \frac{1}{4}$ = ______
10. 4890 = ______ tens
11. Halve 51. ______
12. If you bought three bread rolls at 60c each, how much change would you have from $5.00? ______
13. What is the time if the long hand is on 11 and the short hand is between 3 and 4? ______
14. How many total days are in November and December? ______ days
15. Rotate 90° clockwise.

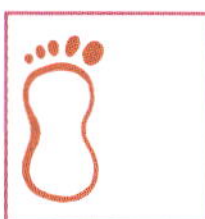

16. Tai needs 2500mL of springwater for the coffee maker. Colour that amount.

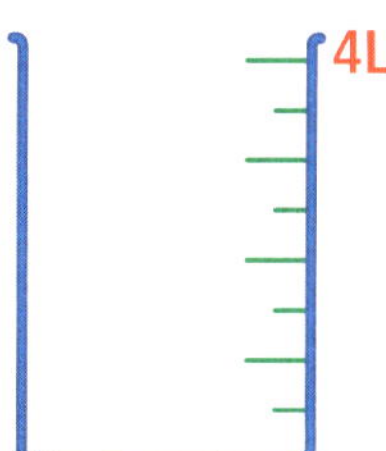

17. Draw ↓ to show which way the balance will tip.

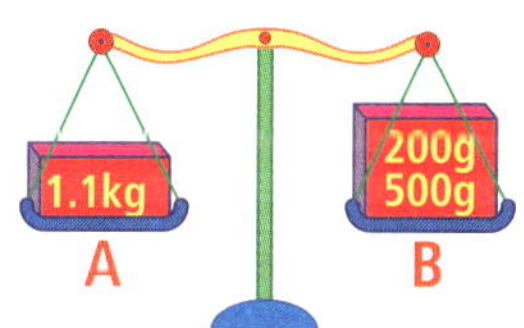

18. Look back at Thursday's spinner. What is the probability of a prime number? ______

Week 29

Monday

1. Each small cube is 1cm^3. What is the volume of the large cube?

_____ cm^3

2. $n \times 7 = 21$, so n = _____.
3. Halve 9. _____
4. What is the length of:

a? _____ b? _____

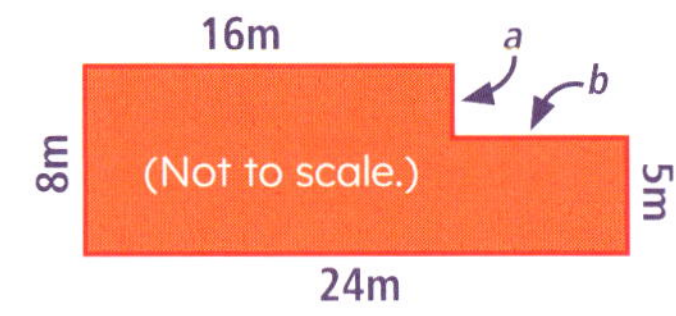

5. $10\frac{1}{6} - \frac{5}{6}$ = _____
6. 800 ÷ 5 = _____
7. 10:32 18 minutes later 
8. 20 + 5 + 0.5 + 0.05 = _____
9. How many more green balloons were popped than red balloons? _____

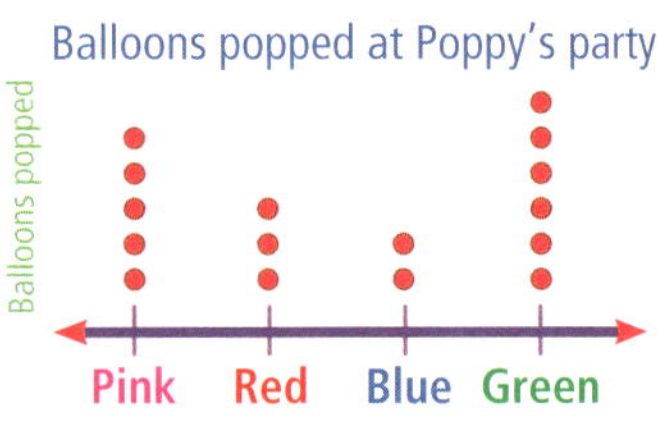

10. What is the value of the 3 in 4.023?

_____

11. 400 – 17 = _____
12. 6 × 6 = 4 × _____
13. $4\frac{1}{3} - \frac{2}{3}$ = _____
14. Two Olympic freestyle swimmers swam 100m. Their times were 49.9 secs and 50.0 secs.

What was the time difference?

_____

15. 63 ÷ 9 = 14 ÷ _____

Tuesday

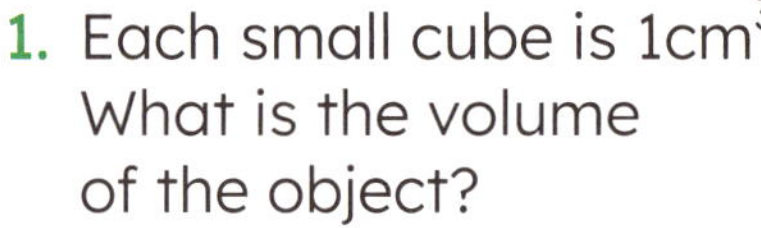

1. Each small cube is 1cm^3. What is the volume of the object?

_____ cm^3

2. What time will the next bus be due if the intervals are equal?

Departures
9:05 am
9:20 am
9:35 am

_____

3. (a) 4 × 5 = _____ (b) 14 × 5 = _____
4. Tick the quotient of 8.

40 ÷ 5 = 8 ☐ 24 ÷ 8 = 3 ☐

5. 101 – 7 = _____

1001 – 7 = _____

6. Which is 3D, a triangle or a pyramid?

_____

7. 81 ÷ 9 = 27 ÷ 3 = _____
8. Match each decimal with its fraction.

0.12	$1\frac{2}{10}$
1.2	$\frac{12}{1000}$
0.012	$\frac{12}{100}$

9. 40 + 4 + 0.2 + 0.09 = _____
10. What is the cost of buying 10kg of prawns at $20.00 per kg?

_____

11. 5 × 0.8 = _____
12. Draw a reflection of:

13. $\frac{3}{4} = \frac{\square}{8}$
14. Write $\frac{1}{2}$ as a decimal. _____
15. During a race, Macy swam 49.998 secs, while Kyra swam 49.988 secs. What was the time difference?

_____

Wednesday

1. The volume of this object is ______ cm³.

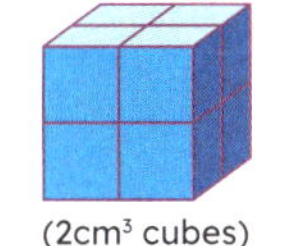

2. 9.07 = 9 + 0.______

3. Draw ↓ to show the heavier side.

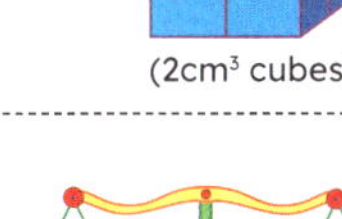

4. What number is after 529 899?

______

5. 24 ÷ 6 = ______

6. 2.59 × 10 = ______

7. Draw the line of symmetry.

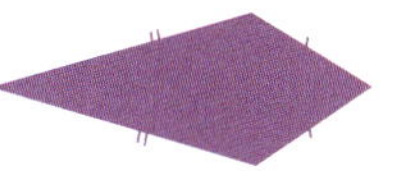

8. Complete the number line in decimal form.

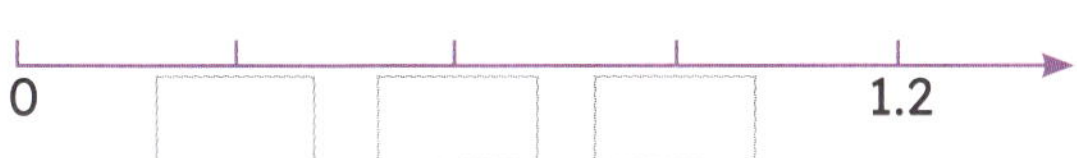

9. The factors of 48 are:

1, ______, ______, ______, ______, ______, ______, 16, 24, 48

10. How many (50c coin) make up $6.50?

______

11. How many cubes make up Tower E?

______

12. $0.5 = \frac{\square}{10}$

13. In 314 200, what is the value of the 3?

______

14. Rank the places by their finishing times.

Race time	Rank
51.010 sec	
51.001 sec	
51.101 sec	

15. 3245 = ______ tens ______ ones

Thursday

1. What is the volume of this object?

______ cm³

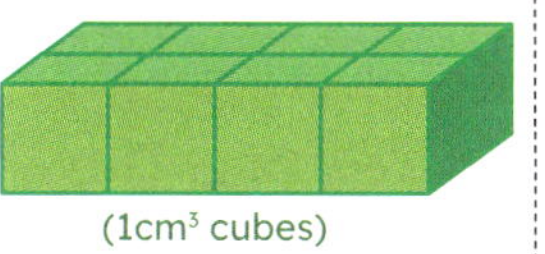

2. 5.46 = 5 + 0.______ + 0.______

3. $2\frac{1}{4} - \frac{3}{4}$ = ______

4. Does the letter Y have a vertical or horizontal line of symmetry?

______

5. Which two spinners have the same chance of landing on a 3?

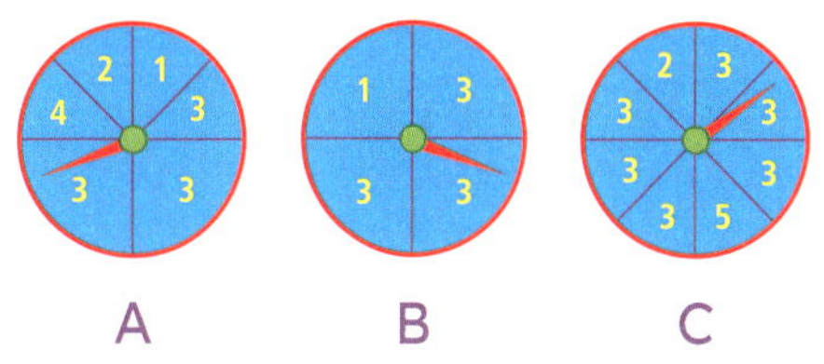

A and B ☐ A and C ☐ B and C ☐

6. What is the area of a 6 by 8 grid? ______

7. 1.1 × 10 = ______

8. Write one million, one thousand and ten as a numeral.

______

9. 10 000, ______, ______, ______, 9996

10. How many weeks are in one year? ______

11. 700 × 10 = ______

12. Which is 2D, a circle or a sphere?

______

13. 

14. Write in ascending order.

2.1 2.01 2.11 2.0 2.001

______ ______ ______ ______ ______

15. Draw a reflection of:

Week 29

Problem-solving

Week 29

The triplets were arguing about the volume of a puzzle cube. They knew that the area of each of the square faces on the cube was $9cm^2$. (= $1cm^3$)

Jarli said the volume of the cube was is $26cm^3$, Marli said $9cm^3$, and Yindi said $27cm^3$.

Let's find out which of the triplets worked out the correct volume.

Read the question again. **Think** about the information. Underline the important words.

Tick the strategy you will use to work out the answer:

- estimate and check ☐
- look for patterns ☐
- draw a diagram or picture ☐
- construct a table or graph ☐
- use materials ☐
- act it out ☐
- work backwards ☐
- something else. ☐

Solve it:

Reflect on the question and answer.

Check it. Circle another strategy on the list to work out the answer.

Show it:

Friday Review

1. Circle the multiples of 6.

   32   36   56   8   54

2. 300 – 17 = ______

3. In 374 655, the value of the 3 is:

   - 3 × 1 ☐
   - 3 × 10 ☐
   - 3 × 100 ☐
   - 3 × 100 000 ☐

4. 38 + 76 = ______

5. What is the value of the 7 in 2.097?

   ______

6. 8 × 0.3 = ______

7. Write one million and one hundred as a numeral.

   ______

8. 32 ÷ 8 = ______

9. How many [50c coin] make up $8.50?

   ______

10. $5\frac{1}{5} - \frac{4}{5} =$ ______

11. 3, 9, 27, 81, ______

12. How many cubes make up the volume of this object?

    ______

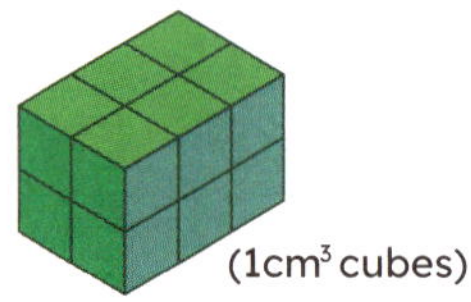

($1cm^3$ cubes)

13. Rank the places.

Race time	Rank
59.094 sec	
59.009 sec	
59.109 sec	

14. Which is 2D, a cube or a square?

    ______

15. 

    17 minutes later

16. In Monday's dot plot, which colour balloon was popped the least?

    ______

17. When is the next bus, if the intervals remain the same?

    ______

Departures
9:05 am
9:20 am
9:35 am

18. Which spinners have the same chance of spinning a 2?

A

B

C

- A and B ☐
- B and C ☐
- A and C ☐
- A, B, and C ☐

Monday

1. Which city has the same time?

   4:00 pm **Naarm / Melbourne**

   Boorloo / Perth ☐
   nipaluna / Hobart ☐
   Gulumoerrgin / Darwin ☐

2. If the short hand of a clock is between 12 and 1, and the long hand is on 11, what is the time?

   ____

3. Write three hundred and fifty thousand as a numeral.

   ____

4. Double 45. ________

5. 18:35 25 minutes before 

6. $2 - \frac{2}{5} =$ ________

7. 1ha = ________ m^2

8. Rotate this pattern 270° clockwise. 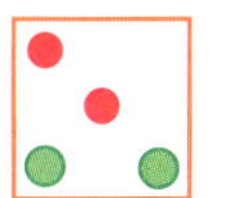 

9. Which pair contains 2 prime numbers?

   9, 15 ☐ 21, 31 ☐ 23, 31 ☐ 23, 25 ☐

10. 3, 23, 43, 63, ________

11. 40 × 6 = ________

    39 × 6 = 234

    38 × 6 = ________

12. How many cubes make up the volume of this object?

    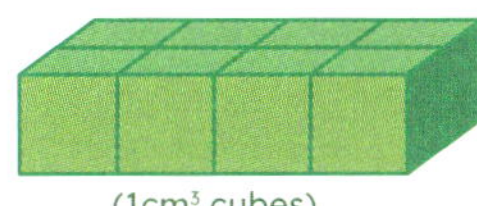

    ________ cubes

13. 6 × 1000 = ________

14. Write the fractions on the number line.

    $\frac{1}{2}$ $\frac{1}{3}$ $\frac{4}{5}$

    0 ☐ ☐ ☐ 1

15. 9 × 8 = 18 × 4 = 36 × ________ = ________

Tuesday

Week 30

1. How many minutes is a quarter of an hour? ________

2. Add one-third of an hour to the 24-hour time.

   20:25 

3. Arrange 6, 4, 9, and 0 to create the lowest whole number.

   ____

4. $4\frac{4}{5} + \frac{4}{5} =$ ________

5. 90cm = 0.________m

6. Zac recorded 12 throws in a dice game: 3, 7, 7, 6, 8, 4, 12, 7, 8, 5, 6, 11.

   (a) Order the numbers from smallest.

   ____

   (b) Find the mode.

   ____

7. Round 10.75 to the nearest whole number. ________

8. 60 + 0.03 + 0.9 + 2 = ________

9. 900 − 50 = ________

10. If cut, what 2D shape would you see on the cross-section?

    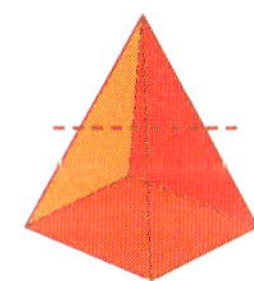

    ____

11. 108 ÷ 12 = 54 ÷ 6 = 27 ÷ 3 = ________

12. Round 6.739 to the nearest hundredth.

    ____

13. Match each name with its irregular shape.

    pentagon

    octagon

14. 9.2 + 0.9 = ________

15. How many degrees does a circle have?

    ____

Week 30

Wednesday

1. What is the date one fortnight after 16 April?

 ____

2. Your money jar contains 6 × \$1, 4 × \$2, and 3 × 50c.

   What is the total amount? __________

3. How many laps of a 50-metre pool have to be swum to complete 1700m?

 ____

4. (a) The acute angle is __________.

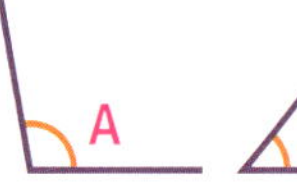

   (b) The obtuse angle is __________.

5. Halve $\frac{1}{4}$. __________

6. Rotate this pattern 180°.

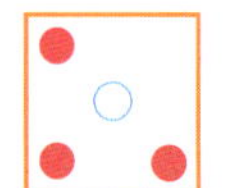

7. Luke stacked ten 10c coins and made a:

   cube. ☐ decagon. ☐

   cylinder. ☐ cone. ☐

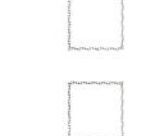

8. 10 000 – 2500 = __________

9. What type of triangle is this?

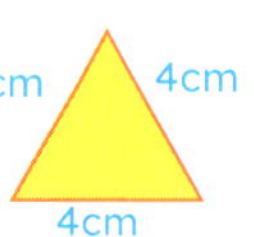

(Not to scale.)

 ____

10. 8.36 × __________ = 836

11. 4.3 + 0.8 = __________

12. n × 8 = 4, so n = __________.

13. 2, 10, 50, __________, 1250

A farmer planned some tree planting.

A		A		A
A	M	A	M	A
A	M	A	M	A
A	M	A	M	A
A		A		A

Scale

☐ = 25m^2

A = 2 almond trees

M = 3 macadamia trees

14. What is the total area? __________

15. How many almond trees are being planted? __________

Thursday

1. How many days in 48 hours?

 ____

2. (25 ÷ 5) × (30 ÷ 6) = __________

3. 15 ÷ 3 = 30 ÷ 6 = 60 ÷ 12 = __________

4. There is a 20-minute interval. How long is the show at Sydney Opera House?

   SYDNEY OPERA HOUSE SHOW
   10 am – 2 pm

 ____

5. The factors of 81 are:

   1, ______, ______, ______, 81

6. Alvin has \$3.45 and wants to buy an ice cream at \$4.20.

   Alvin is short by __________.

7. 600m + 900m = __________m

8. Which shape is divided into thirds?

 ____

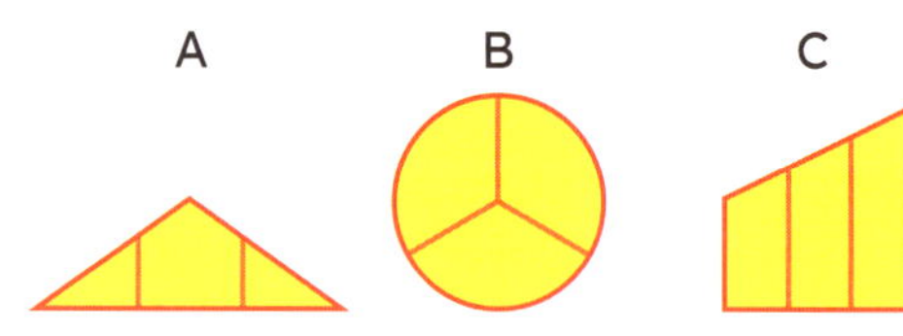

9. How many degrees in a straight line?

 ____

10. 0.08 + 0.70 = __________

11. Join the dots to create two cubes.

12. 0.7 = $\frac{7}{1}$ ☐ $\frac{7}{10}$ ☐ $\frac{7}{100}$ ☐ $\frac{7}{1000}$ ☐

13. Match each decimal with its fraction.

0.035	$1\frac{35}{100}$
1.35	$\frac{35}{100}$
0.35	$\frac{35}{1000}$

14. 9730 = __________ tens

15. $8\frac{1}{3} - \frac{2}{3}$ = __________

Problem-solving

Ethan wants to phone relatives to thank them for some birthday presents. Granny Annie lives in Darwin and Uncle Bob lives in Perth. Ethan lives in Sydney, where it is 3:30 pm (non-daylight saving).

Let's find out what time it will be (in 12- and 24-hour time) in Darwin and Perth if Ethan calls his relatives now.

Read the question again. **Think** about the information. Underline the important words.

Tick the strategy you will use to work out the answer:

- estimate and check ☐
- look for patterns ☐
- draw a diagram or picture ☐
- construct a table or graph ☐
- use materials ☐
- act it out ☐
- work backwards ☐
- something else. ☐

Solve it:

Reflect on the question and answer.

Check it. Circle another strategy on the list to work out the answer.

Show it:

Friday Review

Week 30

1. $5\frac{3}{5} + \frac{4}{5} =$ ______
2. Which letter on the number line shows $\frac{3}{4}$? ______

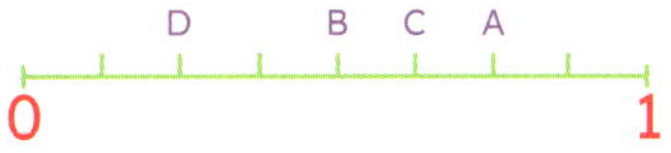

   (number line from 0 to 1 with points D, B, C, A)
3. $7 \times 0 =$ ______
4. $(30 \div 10) \times (42 \div 6)$ = ______
5. If you had two \$1 coins, three \$2 coins, and eight 20c coins, what is the total? ______
6. Round 3.259 to the nearest hundredth. ______
7. $n \times 9 = 18$, so n = ______.
8. Match each decimal with its fraction.

0.29	$\frac{2}{100} + \frac{9}{1000}$
2.9	$2\frac{9}{10}$
0.029	$\frac{2}{10} + \frac{9}{100}$

9. $56 \div 8 = 28 \div 4 =$ $14 \div$ ______ $=$ ______
10. $50 + 0.03 + 0.9 + 5$ = ______
11. How many 50m laps would you have completed if you swam 2000m? ______
12. 3820 = ______ tens
13. Rotate 90° clockwise.

14. This angle is: acute. ☐ obtuse. ☐ 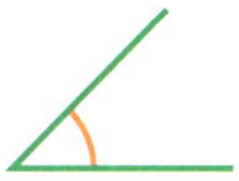
15. What is the date one fortnight from 17 July? ______
16. Draw the corresponding times (non-daylight saving).

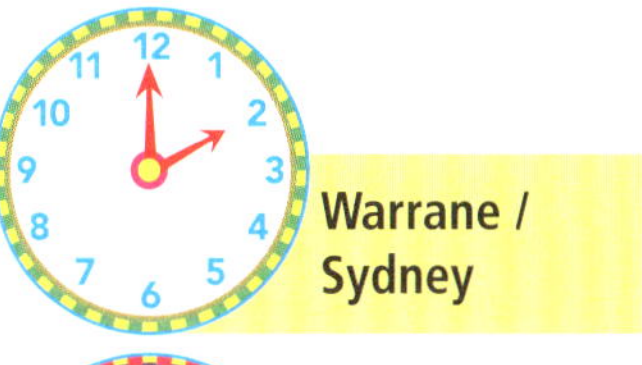

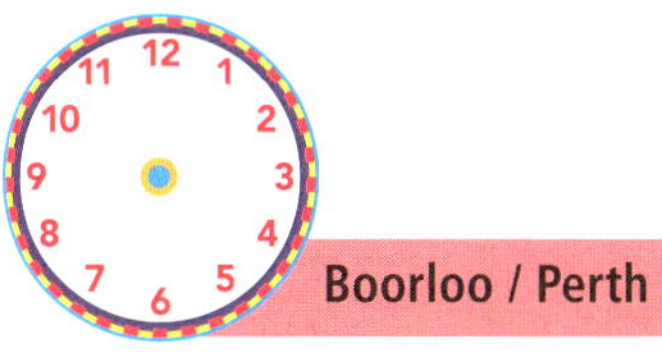

17. 2760 = ______ tens
18. How many cubes make up the volume of this object? ______

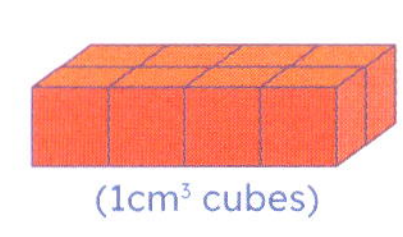
(1cm³ cubes)

Week 31

Monday

1. (a) Double 0.9. ________

   (b) Halve 0.9. ________

2. 790 + 310 = ________

3. A sea container 20m × 2m × 3m was used to transport new TVs in boxes 2m × 2m × 1m. How many TV boxes will fill this container?

   ____

4. 9.7, 9.8, 9.9, ________

5. Which shape(s) is/are divided into quarters?

   ____

   A 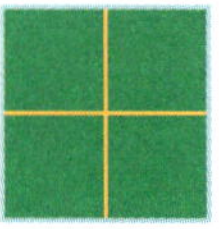 B 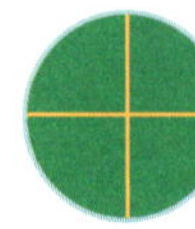 C 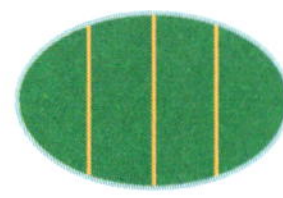 D 

6. $8\frac{3}{5} + \frac{4}{5} =$ ________

7. 1.4 × 10 = ________

8. 715 396 – ________ = 700 000

9. This painting was bought for $300 000 and sold at 10% profit.

   Profit = $________

10. 0.2 < 1 true ☐ false ☐

11. 25 × 32 = ________ × 8

12. 9789 × 10 = ________

13. This is a net of a

    ____

    ________.

14. 2 × (3 × 19) = ________

15. The best buy is:

    three tulips for $19.50 ☐

    two tulips for $15 ☐

    one tulip for $6.95 ☐

Tuesday

1. (a) Double 191. ________

   (b) Halve 191. ________

2. 3.7 + 0.4 = ________

3. A hexagon has rotational symmetry to the order of ________.

4. 9 + 0.002 + 0.2 + 0.08 = ________

5. 750 ÷ 5 = ________

6. Draw each line of symmetry. (Use the centre dot as a guide.)

   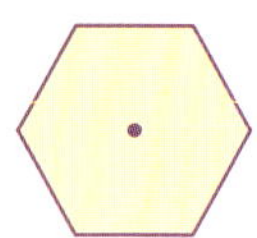

7. $\frac{4}{20} = \frac{\square}{10}$

8. What is the time difference? ________

9. 900kg + 200kg = ________t

10. (6 + 4) × (4 × 3) = ________

11. Label the angles in this shape.

    A = acute

    B = obtuse

    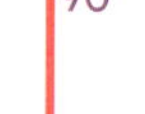

12. 100m freestyle times: Alicia 58.9 secs, Jessica 59.2 secs.

    What is the time difference? ________

13. $\frac{11}{3} = 3\overline{)11} = 11 \div 3 =$

    $3\frac{2}{11}$ ☐ $3\frac{2}{3}$ ☐ $\frac{3}{11}$ ☐

14. 1 × 24 = ________ 2 × ________ = 24

    3 × ________ = 24 4 × ________ = 24

15. Packets of chips are packed in boxes. Calculate the total number of packets in boxes A, B, and C.

    ____

Wednesday

1. (a) Double 135. ________

   (b) Halve 135. ________

2. $8\frac{5}{6} + \frac{3}{6} =$ ________

3. 500 – 23 = ________

4. 0.07 + 0.4 + 20 + 4 = ________

5. Draw ↓ to show the heavier side.

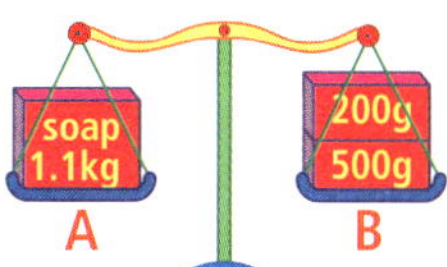

6. $\frac{1}{4}$ of 1L = ________mL

7. A truck has a gross mass of 24t. After unloading, the truck's mass is a third less. What was the net mass of the load?

   ____

8. 400g = ________kg

9. This is an irregular ________.

10. How many students were surveyed?

    ____

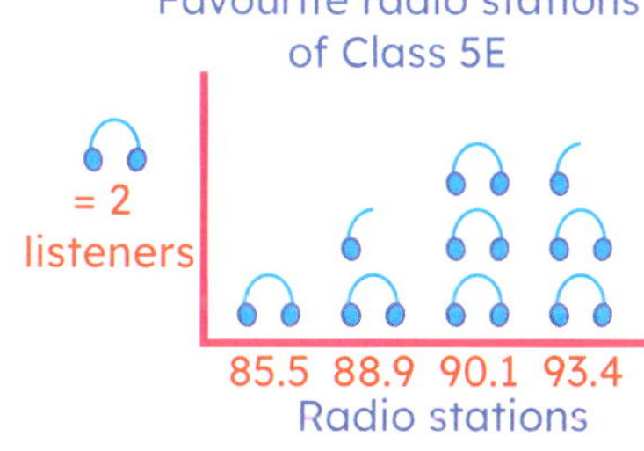

11. $3.25 = \frac{\square}{100}$

12. 0.5 × 100 = ________

13. What is the probability of landing on an odd number?

    ____

14. What is the cost of buying 100kg of lemons at $1.50 per kg?

    ____

15. Write $\frac{1}{5}$ as a decimal. ________

Thursday

Week 31

1. (a) Double 1.2. ________

   (b) Halve 1.2. ________

2. 780 + 320 = ________

3. Marty swam 100m in 56.309, while Alex swam it in 56.298. Who was faster and by what time?

   ____

4. Balance the scales.

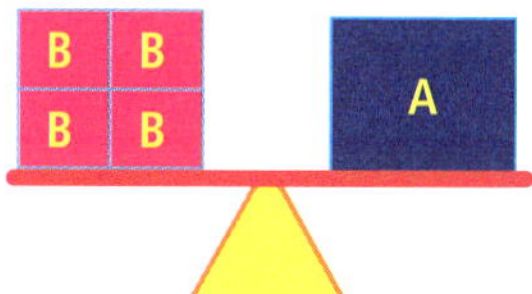

   A = 1t

   B = ____.____ t

5. $1 - \frac{1}{3} =$ ________

6. 9 × 69 = ________

7. A fast train travelled 900km in 2 hours.

   Its speed was ________km/h.

8. Olivia flipped a coin and recorded the results. What is the chance of the next flip landing on heads?

Heads	Tails
𝍸	

   ____

9. If a square building has a perimeter of 120m, what is the length of one of its sides? ________m

10. The HCF of 12 and 16 is ________.

11. Colour the jug to show 300mL.

12. 0.78 + 0.10 = ________

13. Tick the quotient of 3.

14. Write nine million, nine hundred and nine thousand, and nineteen as a numeral.

    ____

15. What is the area of the office floor?

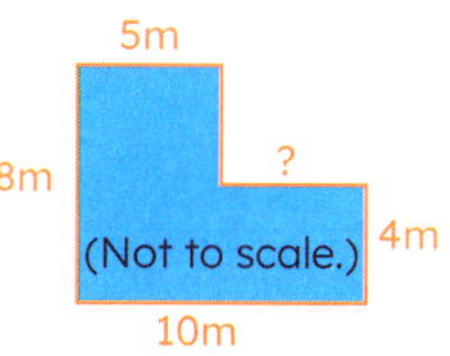

    ____

Week 31

Problem-solving

Coen has the following ingredients to make 24 cupcakes:

325g butter	200g sugar
3 eggs	250g icing sugar
180g flour	2.5 tsp vanilla extract
3.5 tbsp milk	

However, they want to make 48 cupcakes.

Let's find out how much of each ingredient Coen will need. (Hint: Use digital tools to solve.)

Read the question again. **Think** about the information. Underline the important words.

Tick the strategy you will use to work out the answer:

- estimate and check ☐
- look for patterns ☐
- draw a diagram or picture ☐
- construct a table or graph ☐
- use materials ☐
- act it out ☐
- work backwards ☐
- something else. ☐

Solve it:

Reflect on the question and answer.

Check it. Circle another strategy on the list to work out the answer.

Show it:

Friday Review

1. 840 + 190 = ______
2. 89 × 7 = ______
3. $3\frac{1}{8} - \frac{5}{8}$ = ______
4. In Wednesday's graph, how many more students liked 90.1 than 88.9? ______.
5. 0.7, 0.8, 0.9, ______
6. Write $\frac{1}{5}$ as a decimal. ______
7. 10 000 – 2900 = ______
8. (a) Double 153. ______

   (b) Halve 153. ______
9. In a jar of seven red, eight blue, and three yellow pens, what is the probability of choosing a yellow pen? ______
10. 28 ÷ 7 = ______
11. $\frac{14}{6}$ = $6\overline{)14}$ =

    14 ÷ 6 =

    2 ☐ $2\frac{2}{14}$ ☐ $2\frac{2}{6}$ ☐
12. Write ten million one hundred and ten as a numeral. ______
13. 2.65 × 10 = ______
14. You paid $900 000 for a painting and sold it at 20% profit.

    Profit = $______
15. 900mL = 0.______L
16. A is: 2D ☐ 3D ☐

    B is: 2D ☐ 3D ☐

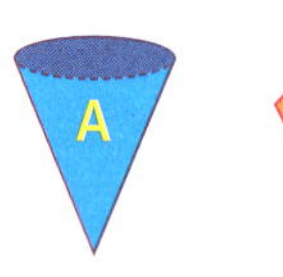

17. What is the area of this shape?

    ______m^2

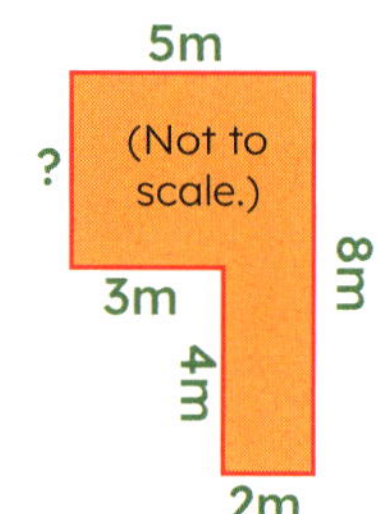

18. After completing 20 coin flips, there were four more heads than tails. Complete the tally to show the final results.

Heads	Tails
II	𝍸

St P

Monday

1. Round 87 050 to the nearest hundred. ________

2. 0.3 + 0.05 + 50 + 2 = ________

3. If you double the length of $\overline{XY}$, what is its new length?

________cm

X ———————— Y

4. Join the dots to create two stacked cubes.

5. 20 + 300 000 + 40 000 + 800 + 7

= ________

6. 2 × 7 = 14, 4 × 7 = 28, 8 × 7 = ________

7. 4.45 × 10 = ________

8. 0.4t = ________kg

9. $6 - 2\frac{1}{2}$ = ________

10. 99 999 + 100 = ________

11. $\frac{2}{3} < \frac{4}{5}$ true ☐ false ☐

12. What is the volume? ________

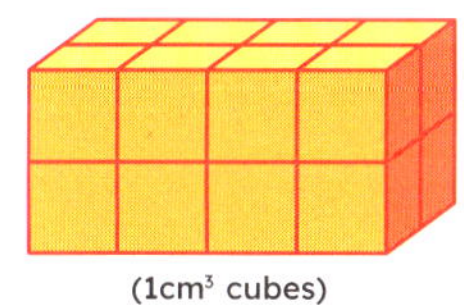

(1cm³ cubes)

13. Write an equivalent fraction of $\frac{4}{6}$.

____

14. Circle the multiples of four.

102 100 46 44 36 28

15. 61 297 – ________ = 60 000

Tuesday

1. Round 73 909 to the nearest ten thousand. ________

2. 2 × 8 = 16

4 × 8 = 32

8 × 8 = ________

3. Isla had 27 coloured pencils, of which three were red. What fraction of pencils were not red?

________ simplified to ________

4. Convert $1 + \frac{7}{100}$ to a decimal. ________

5. 600mm = ________m

6. 4.7 × 100 = ________

7. $9 - 3\frac{1}{2}$ = ________

8. What is the cost of 100kg of baked beans at $2.50 per kg? ________

9. Complete the timetable with 15 minute departure intervals.

Departures
8:03 am
8:18 am
________ am
________ am

10. 18 ÷ 6 = ________

11. 0.27 + 0.009 = ________

12. Halve $\frac{1}{2}$. ________

13. Share $25.00 equally among 10 people.

____

14. 4.2 – 0.4 = ________

15. 856 ÷ 4

800 ÷ 4 = ________

40 ÷ 4 = ________

16 ÷ 4 = ________

Total = ________

Week 32

Wednesday

Week 32

1. Round 7.047 to the nearest tenth. ________

2. Draw the left-side view.

3. To swim a distance of 1km, how many laps of a 25m pool are needed?

____

4. Complete the timetable with 14 minute departure intervals.

Departures
7:07 am
________ am
________ am

5. 0.8L = ________mL

6. 1 000 000 + 10 000 + 11

= ________

7. Write five million and fifty as a numeral.

____

8. 0.2 > 1 true ☐ false ☐

9. 3.2 – 0.5 = ________

10. $5 - \frac{3}{8} =$ ________

11. Convert $1 + \frac{3}{10} + \frac{8}{10}$ to a decimal.

____

12. (a) Start at A. Colour (D,2), (C,2), and (B,2), and label (A,2) as B.

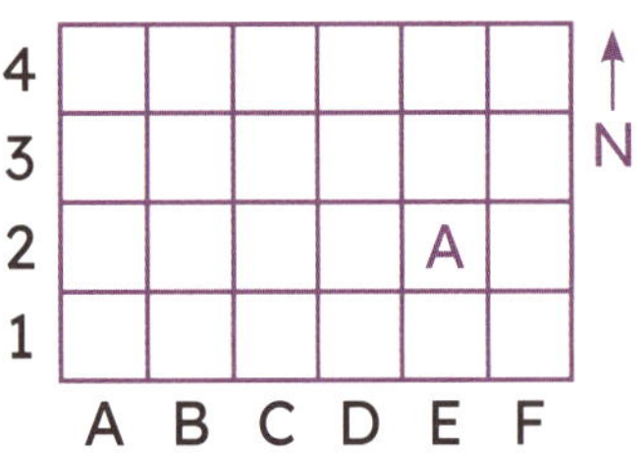

(b) What is the direction from A to B?

____

13. ∠ BCD = 60°, so ∠ ACD = ________°.

D
60°
B C A

14. Which angle is obtuse? ________

15. Share $45.00 equally among 10 people.

____

Thursday

1. Round 4.057 to the nearest hundredth. ________

2. 59 ☐ 10 = 5.9

3. Which is a net of a cube?

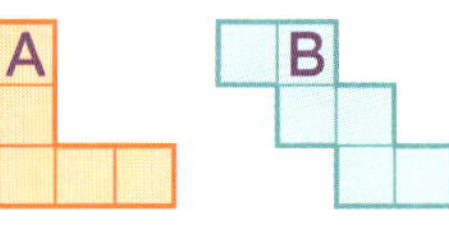

A ☐ B ☐

4. 0.4m = ________cm

5. 0.02 + 0.5 + 30 + 6 = ________

6. 3 × 7 = 21

6 × 7 = 42

12 × 7 = ________

7. odd – even = ________

8. What is the probability of landing on a prime number?

________ in ________.

9. 2, 6, 12, 36, 72, 216, ________

10. Write $\frac{12}{15}$ in its simplest form. ________

11. $\frac{1}{8} > \frac{1}{3}$ true ☐ false ☐

12. 410 + 790 = ________

13. Amy read 450 pages of a book in five hours. What is the average number of pages read per hour?

____

14. Which month was the coldest?

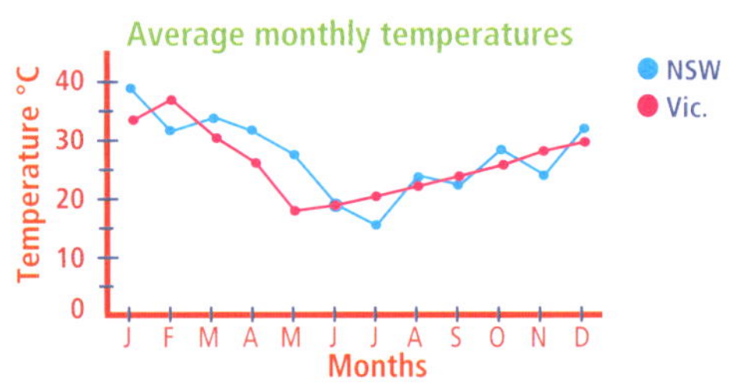

15. 3 × 24 = ________

Problem-solving

350 students need transporting to the local pool for swimming lessons. A bus seats 54 passengers.

Let's find out how many buses are needed and how many spare seats there will be.

Read the question again. **Think** about the information. Underline the important words.

Tick the strategy you will use to work out the answer:

- estimate and check ☐
- look for patterns ☐
- draw a diagram or picture ☐
- construct a table or graph ☐
- use materials ☐
- act it out ☐
- work backwards ☐
- something else. ☐

Solve it:

Reflect on the question and answer.

Check it. Circle another strategy on the list to work out the answer.

Show it:

Friday Review

1. 27 ÷ 9 = ______
2. 429 768 – ______ = 400 000
3. Write $\frac{18}{24}$ in its simplest form. ______
4. 4 × 7 = 28
   8 × 7 = 56
   16 × 7 = ______
5. $4\overline{)328}$ = ______
6. Halve $\frac{1}{4}$. ______
7. What are the chances of selecting a prime number from a hat containing the numbers 1–10?
   ______ out of ______
8. $4\frac{3}{4} + \frac{3}{4}$ = ______
9. Round 83 080 to the nearest ten thousand. ______
10. Which two prime numbers add to 19? ______
11. $20.00 – $12.50 = ______
12. The best buy is:
    Four roses for $28.74 ☐
    Two roses for $16.50 ☐
13. 21 ☐ 10 = 2.1
14. 0.8m = ______ mm
15. ∠ DBC = 40°, so ∠ ABC = ______°.

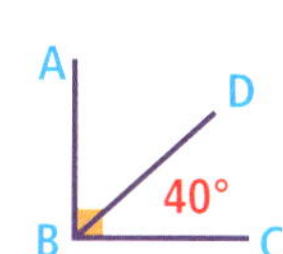

16. This is a net of a ______.

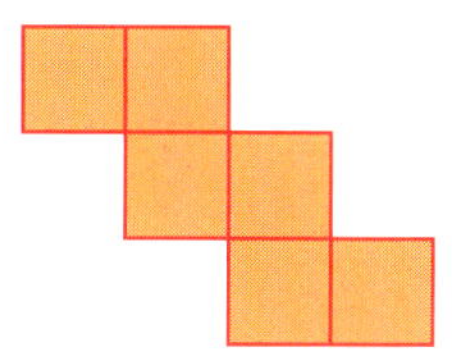

17. Look back at Thursday's graph. Which months were hotter in Victoria than New South Wales?
    ______
18. In which month were the average temperatures equal for both states?
    ______

A

St

Week 33

Monday

1. To land on a 2, spinner(s):

A has a better chance. ☐

B has a better chance. ☐

A and B have an equal chance. ☐

2. Add four tens to three lots of six. ______
3. 3 × 6 = 18 6 × 6 = 36 12 × 6 = ______
4. $\frac{6}{10} + \frac{4}{100} = 0.$______
5. \$10.00 – \$4.50 = ______
6. 2.73 ÷ 10 = ______
7. There are seven rows of eight chairs. How many chairs in total? (Write as a number sentence.)

______

8. Write $\frac{8}{12}$ in its simplest form. ______
9. What is the value of the 2 in 2 000 000? ______
10. 2.2km = ______m
11. Rotate 90° anticlockwise.

12. Each box holds different-sized bottles of milk. How many bottles are in each box?

A = ______ bottles B = ______ bottles

C = ______ bottles D = ______ bottles

13. What is the product of 4 and 8? ______
14. 2800 ÷ 5 = ______
15. Colour (D,1), (C,2), (B,3), and (A,4). What direction is it from:

(a) P to T? ______

(b) T to P? ______

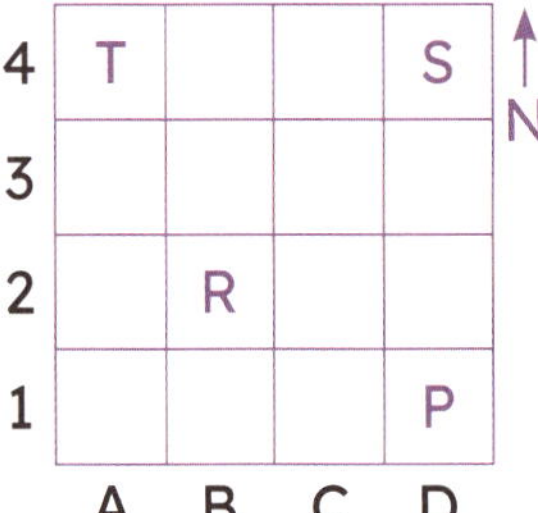

Tuesday

1. If you wanted to land on an even number, spinner(s):

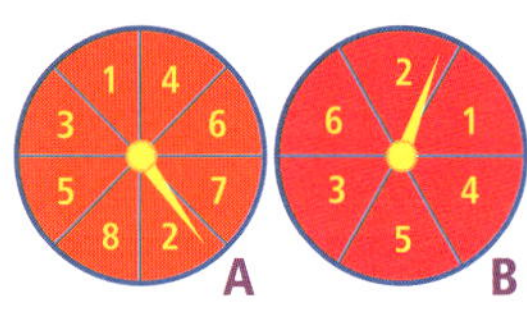

A has a better chance. ☐

B has a better chance. ☐

A and B have an equal chance. ☐

2. 6.072 × 1000 = ______
3. Draw a line on the cube so a cut creates a triangular prism.

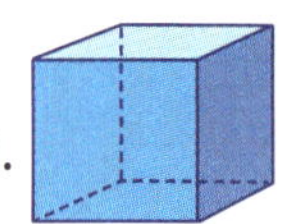

4. $1 - \frac{2}{10} =$ ______
5. Count the vertices.

A = ______

B = ______

C = ______

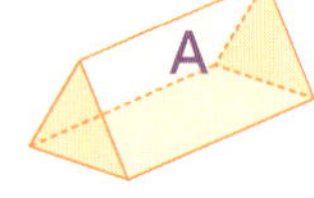
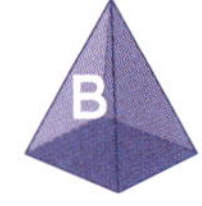
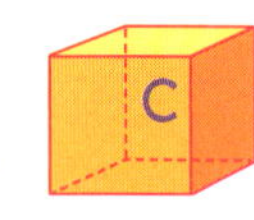

6. Share \$95.00 equally among ten people.

______

7. 15, 30, 45, ______, ______, 90
8. In 246 000, the 2 = 2 × ______
9. Match each name to its shape.

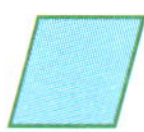

rhombus kite square

10. (3 × 9) ÷ (9 ÷ 1) = ______
11. Round 61 092 to the nearest hundred. ______
12. 380 + 720 = ______
13. What is the length of side A?

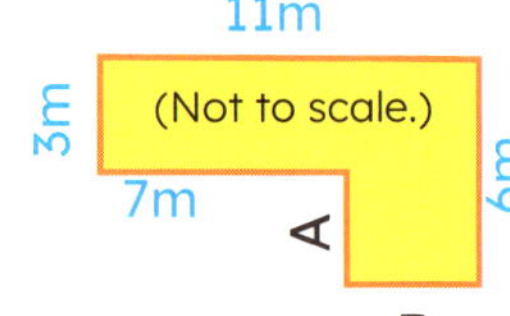

______

14. What is the length of side B?

______

15. What is the perimeter? ______

Wednesday

1. Which statement is correct?

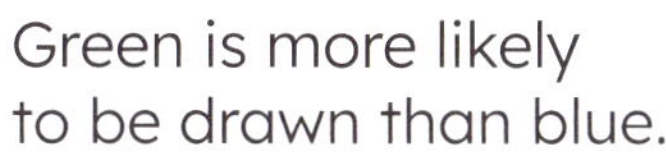

Blue is more likely to be drawn than green. ☐

Green is more likely to be drawn than blue. ☐

Green and blue are equally likely to be drawn. ☐

2. $\frac{23}{100} + \frac{11}{100} = 0.$______

3. Add eight tens to three lots of eight. ______

4. $\frac{1}{4}$ of 240 = ______

5. $\frac{1}{2} < \frac{1}{3}$ true ☐ false ☐

6. What is the area?

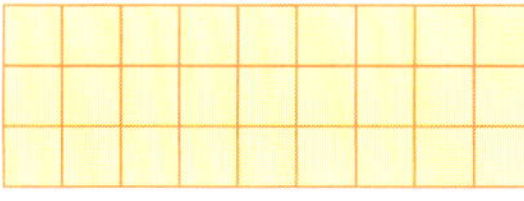

______ squares

7. $2 - \frac{4}{5} =$ ______

8. Which set has 3 symmetrical letters?

A F H ☐ Y Z N ☐ Y A T ☐

T G M ☐ W X Z ☐

9. 1712, 1812, 1912, ______

10. Write these fractions in ascending order.

$\frac{3}{4}$ $\frac{1}{2}$ $\frac{1}{5}$ $\frac{2}{6}$ $\frac{1}{4}$

______ ______ ______ ______ ______

11. $\frac{8}{10} + \frac{14}{100} = \frac{\square}{100}$

12. (4 × 5) + (20 – 8) = ______

13. What is the missing factor of 36?

1, 2, 3, 4, 6, ______, 12, 18, 36

14. Round 2.041 to the nearest hundredth. ______

15.

= $______

Thursday

1. Which statement is correct?

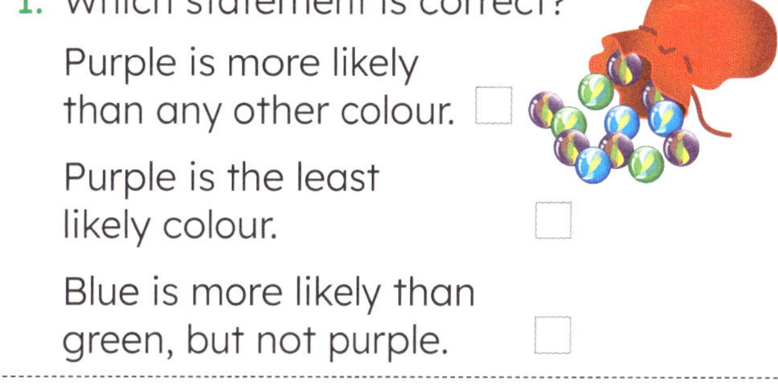

Purple is more likely than any other colour. ☐

Purple is the least likely colour. ☐

Blue is more likely than green, but not purple. ☐

2. 1004 – 7 = ______

3. Write 25% as a fraction other than $\frac{25}{100}$.

______

4. What season is it?

______

5. 3 × 8 = 24 6 × 8 = 48 12 × 8 = ______

6. $\frac{2}{100} + \frac{4}{10} = \frac{\square}{100}$

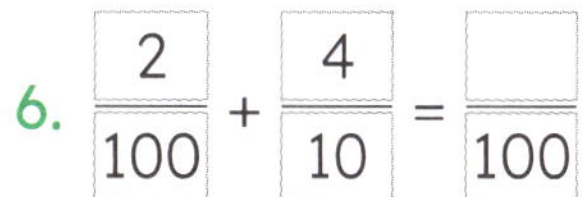

7. $9\overline{)6318}$ = ______

8. This is a net of a

______.

9. Which set has 2 equivalent fractions?

$\frac{1}{5}, \frac{1}{10}$ ☐ $\frac{2}{4}, \frac{4}{8}$ ☐ $\frac{3}{4}, \frac{6}{10}$ ☐ $\frac{1}{2}, \frac{2}{6}$ ☐

10. 50 + 80 = ______

11. A litre of milk costs $1.45. What change would you receive from $2.00? ______

12. $5\overline{)245}$ = ______

13. (a) 1.8 × 10 = ______

(b) 1.8 ÷ 10 = ______

14. 99 099 + 10 000 = ______

15. Order from least to greatest in value.

0.8 1.1 $2\frac{1}{2}$ 2

______ ______ ______ ______

Week 33

Problem-solving

Week 33

There are five bottlebrushes, seven roses, three myrtle wattles, and five strawflowers in a garden.

Let's find out which flowers have the lowest chance, the highest chance, and an even chance of a bee landing on them. Give your answer as a fraction.

Read the question again. **Think** about the information. Underline the important words.

Tick the strategy you will use to work out the answer:

- estimate and check ☐
- look for patterns ☐
- draw a diagram or picture ☐
- construct a table or graph ☐
- use materials ☐
- act it out ☐
- work backwards ☐
- something else. ☐

Solve it:

Reflect on the question and answer.

Check it. Circle another strategy on the list to work out the answer.

Show it:

Friday Review

1. Add eight tens to five lots of nine.

   ______

2. 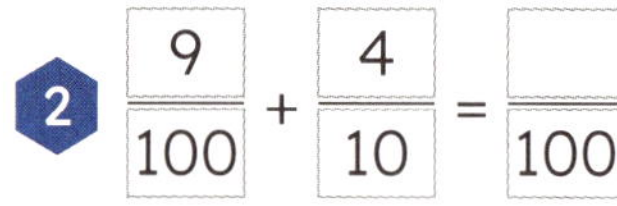

   $\frac{9}{100} + \frac{4}{10} = \frac{\square}{100}$

3. $\frac{5}{10} + \frac{7}{100}$

   = 0.______

4. $3 - \frac{3}{4} =$ ______

5. 1000×0.2

   = ______

6. $5\overline{)265} =$ ______

7. $(3 \times 4) \div (16 \div 4)$

   = ______

8. Write these fractions in ascending order.

   $\frac{1}{2}$ $\frac{1}{4}$ $\frac{2}{3}$ $\frac{4}{5}$ $\frac{3}{4}$

   ____ ____ ____ ____ ____

9. In 682 000, the value of the 6 is

   6 × ______.

10. 1002 − 9 = ______

11. 105, 90, ______, 60, 45

12. 89 009 + 1000

    = ______

13. What is the area?

    ______ squares

    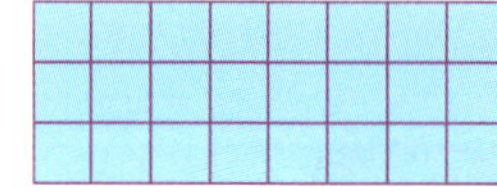

14. Each box has 10 bottles of milk. Write the total litres for each box.

    ______

    ______

    ______

15. Using spinner B from Monday, what is the probability of an odd number?

    ______

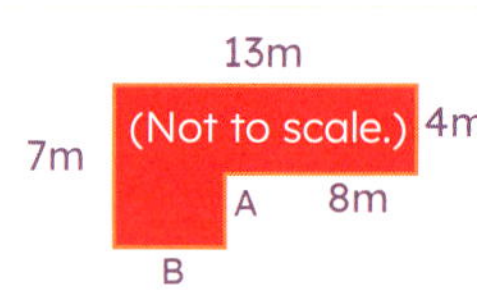

16. What is the length of side A?

    ______m

17. What is the length of side B?

    ______m

18. What is the perimeter of the above shape?

    ______m

P

Monday

1. Draw a reflection of:

2. Add two hundreds to three tens and five lots of eight.

3. Write the next four multiples of 9.

   45, ______, ______, ______, ______

4. Match each city to its time.

Hobart	————	3:00 pm
Melbourne		1:00 pm
Adelaide		3:00 pm
Sydney		2:30 pm
Perth		3:00 pm

5. Write ten million as a numeral. ______

6. 0.97 + 0.10 = ______

7. 1011 – 13 = ______

8. What is 10% interest on $90? ______

9. $0.9 = \frac{\square}{10}$

10. What are the chances of selecting an odd number from a hat containing the numbers 11–20?

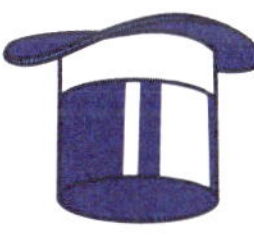

   ______ out of ______

11. 1.1 – 0.1 = ______

12. 15:00 = ______ am ☐ pm ☐

13. 49 × 9 = ______

14. Which is lightest?

   50kg ☐ 5000g ☐ 0.5t ☐

15. 3 400 000 + 600 000 = ______

Tuesday

Week 34

1. Draw a reflection of:

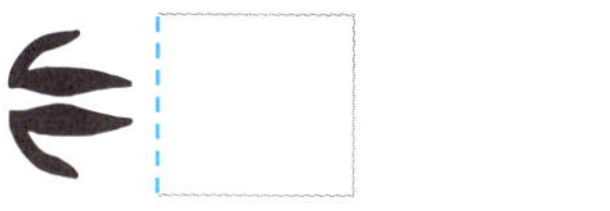

2. 7 × 0.1 = ______

3. There are six bags containing eight oranges. How may oranges are there in total? Write a number sentence.

4. $2 + \frac{2}{10} + \frac{3}{100} =$ ______

5. 8.45km = ______ km and ______ m

6. 14:30 = ______ am ☐ pm ☐

7. What is the sum of 8 and 9? ______

8. 203 ÷ 100 = ______

9. $\frac{19}{100} + \frac{3}{10} = 0.$______

10. What is the area of a 7 by 5 grid?

   ______ squares

11. $4\overline{)220}$ = ______

12. 9090 < 9099 true ☐ false ☐

13. Tim was cutting out a net of a cube. Oops! They cut off one square too many. Which square was cut off?

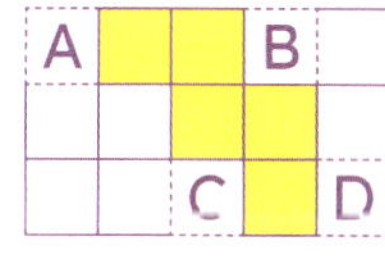

14. If you lived in Warrnambool and drove to Lake Collac, how far would you drive?

15. Enzo, from Tower Hill, drove a V8 ute to Warrnambool. If the ute uses 3L per 2km, how much fuel did Enzo use?

Wednesday

Week 34

1. Draw a reflection of: 

2. 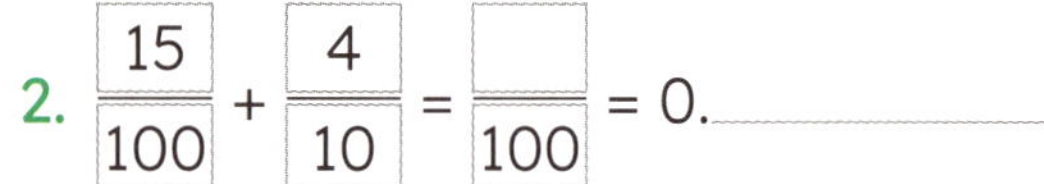 $\frac{15}{100} + \frac{4}{10} = \frac{\square}{100} = 0.$ ______

3. Draw the left-side view.

4. 1.4, 14, ______, 1400

5. Measure the length of $\overline{AK}$. ______ mm

A ———— K

6. Double 11 375. ______

7. Round 67 825 to the nearest thousand. ______

8. 117 ÷ 10 = ______

9. 0.94 + 0.10 = ______

10. How many hours in $1\frac{1}{2}$ days? ______

11. 0.5kg = ______ g

12. (a) ⭐ translated from (3,2) to ______.

(b) 🔴 translated from (4,3) to ______.

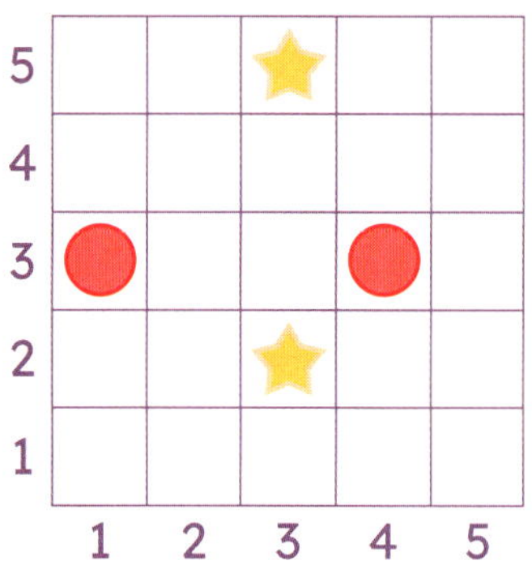

13. 3 × 9 = 27, 6 × 9 = 54, 2 × 9 = ______

14. Draw ↑ to show the lighter side.

15. 69 × 8 = ______

Thursday

1. Draw a reflection of: 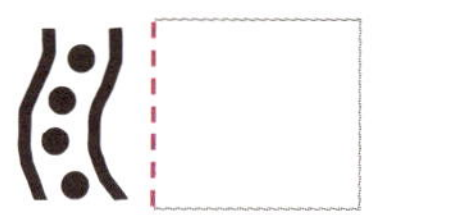

2. Add thirteen tens to four lots of seven. ______

3. Write the next four multiples of 6.

162, ______, ______, ______, ______

4. 420 000 + 80 000 = ______

5. 6.72 × 10 = ______

6. 1.03 – 0.1 = ______

7. In which month did Sam receive the most money?

______

8. odd × odd × even = ______

9. Equilateral triangles have rotational symmetry to the order of three. What is the angle of each rotation?

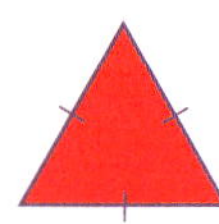

60° ☐ 120° ☐ 80° ☐

10. 9 ÷ 3 = 18 ÷ 6 = ______ ÷ 12 = 3

11. What is the interval between bus departures?

______ minutes

Departures
7:01 am
7:17 am
7:33 am

12. $5\overline{)305}$ = ______

13. 1.1, 11, 110, ______, 11 000

14. $0.8 < \frac{7}{10}$ true ☐ false ☐

15. Write the measurement, in mm, at:

A ______ B ______

C ______ D ______

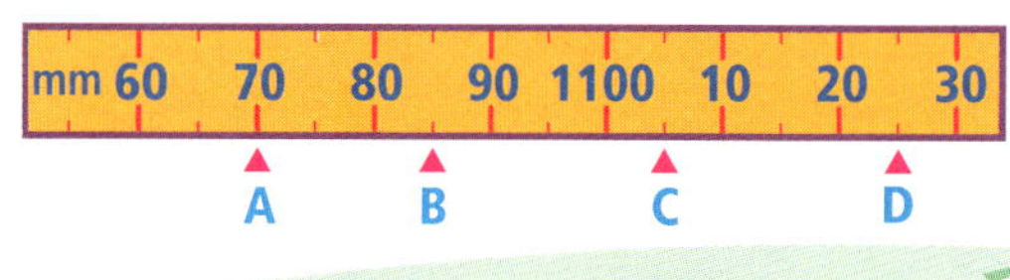

Problem-solving

Aimee was designing a logo for a gymnastics team's outfits. The gymnasts had asked that the logo look the same upside-down and the right way up.

Let's draw an example of what the logo could like.

Read the question again. **Think** about the information. Underline the important words.

Tick the strategy you will use to work out the answer:

- estimate and check ☐
- look for patterns ☐
- draw a diagram or picture ☐
- construct a table or graph ☐
- use materials ☐
- act it out ☐
- work backwards ☐
- something else. ☐

Solve it:

Reflect on the question and answer.

Check it. Circle another strategy on the list to work out the answer.

Show it:

Friday Review

Week 34

1. 8.173×1000 = ______
2. ______ $\div 100 = 2.36$
3. Convert $2 + \frac{4}{10} + \frac{9}{100}$ to a decimal. ______
4. $0.7 < \frac{8}{10}$ true ☐ false ☐
5. $200 \div 2 =$ ______
6. 60 000 + 45 000 = ______
7. $1.05 - 0.1$ = ______
8. Double 16 375. ______
9. Write $\frac{6}{9}$ in its simplest form. ______
10. $0.94 + 0.10$ = ______
11. Which spinner is more likely to land on a 1? ______

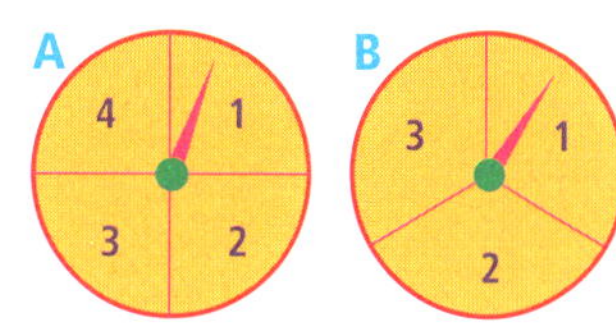

12. 2.57km = ______ km and ______ m
13. A tennis match is broadcast live from Naarm / Melbourne, starting at 9 am. What time will it start on TV live in Tarndanya / Adelaide? ______
14. The measurement at:
    A = ______ mm
    B = ______ mm

15. Draw a reflection of:

16. 

= ______ am ☐ pm ☐
17. Look at Thursday's graph. In which month(s) did Sam not receive pocket money? ______
18. A cube has ______ vertices.

Monday

Week 35

1. Balance the scales.

   C = 200kg

   A = 55kg

   B = ________kg

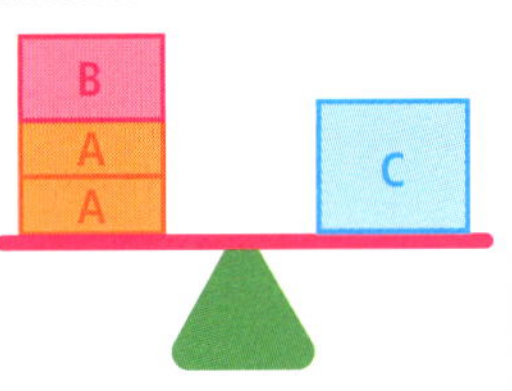

2. You began reading a book at 09:00. By 11:00 you have read 120 out of 300 pages. At what time should you finish the book? **(Sorry, no breaks!)**

   ____

3. 200 ☐ 4 = 100 – 50

4. 0.178 × 1000 = ________

5. Start at Y.

   Move three squares west. Label as Y1.

   Move north one square. Label as Y2.

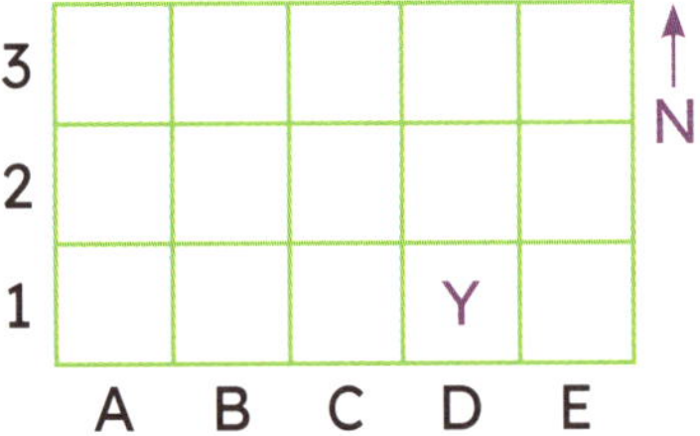

   Move two squares east. Label as Y3.

6. $\frac{1}{5}$ of 1000 = ________

7. 20, 80, 140, ________, 260

8. 86 400 = ________ + 6000 + ________

9. Write $\frac{4}{6}$ in its simplest form. ________

10. 63 ÷ 7 = ________

11. 10 000 – 50 = ________

12. $3 - \frac{2}{5}$ = ________

13. Which equation equals 12 × 8?

    120 + 8 = 128 ☐ 100 – 4 = 96 ☐

    47 + 48 = 95 ☐ 120 – 8 = 112 ☐

14. 18 900 > 18 990 true ☐ false ☐

15. 29 750 + 250 = ________

Tuesday

1. Balance the scales.

   D = 500kg

   C = 150kg

   A + B = ________kg

2. Rotate 270° anticlockwise.

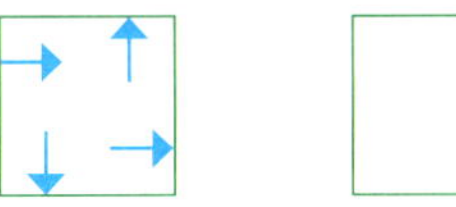

3. 1.4 × 10 000 = ________

4. 2397 ÷ 1000 = ________

5. 24 ÷ 5 = 48 ÷ 10 true ☐ false ☐

6. At the store, potatoes cost $7.50 for 2.5kg. What is the total for 7.5kg of potatoes?

   ____

7. 707 907 – ________ = 700 000

8. 10 × 12 = ________

9. 21 + 19 = ________

10. Stan was cutting out a net for a square-based pyramid. Oops! They cut too much off. Draw in the missing shapes.

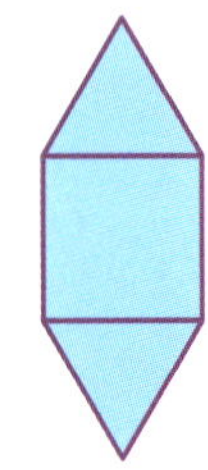

11. In 385 629, the 3 has a value of:

    3 × 100 ☐ 3 × 1000 ☐

    3 × 10 000 ☐ 3 × 100 000 ☐

12. 997 + 8 = ________

13. Complete the timetable to show 15 minute intervals.

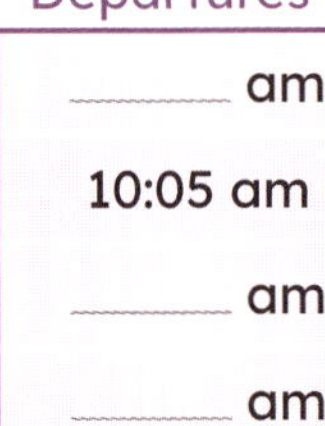

14. 8.5m = ________cm

15. 600 ÷ 5 = ________

Wednesday

1. Balance the scales.

   A = ________kg

   B = 240kg

2. Rotate 360°.

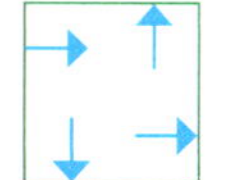

3. $\frac{4}{1000} + \frac{3}{10} = 0.$________

4. Colour the sides parallel to the green side.

5. 9 ☐ 9 = 81

6. A rectangle has an area of $28cm^2$ and its perimeter is 22cm. What is the rectangle's:

   (a) length? ________

   (b) width? ________

7. 400 ÷ 5 = ________

8. $9\frac{1}{4} - \frac{3}{4}$ = ________

9. 7 × 6 = ________

10. A library has six shelves. Each shelf holds 50 books. How many books in total? (Write as a number sentence.)

    ____

11. 1 159 752 – ________________ = 1 000 000

12. Circle the multiples of 7.

    12 18 24 17 21 42

13. 

35 minutes later

14. 999 999 + 9 = ________________

15. *Fold paper in half.* *Cut shape.* *Unfold.*

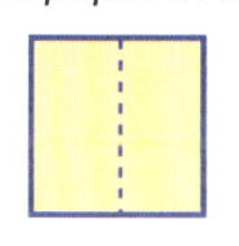
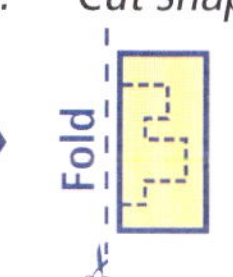

Draw new shape.

Thursday

Week 35

1. Balance the scales.

   A = 120kg

   B = 45kg

   C = ________kg

2. $15 \div 4 = \frac{15}{4}$ = ________ (mixed number)

3. 6 × 1000 = ________

4. 2)1004 = ________

5. 72 ÷ 6 = ________ ÷ 3 = 12

6. Add 100 to 100 000 = ________________

7. Mia was cutting out a net of a cube. They cut off one square by mistake. Which square was cut off?

   ______

8. A rectangle has an area of $28cm^2$ and its perimeter is 32cm. What is the rectangle's:

   (a) length? ________

   (b) width? ________

9. 2.095 ÷ 10 = ________

10. This is an irregular ________________.

11. What is the perimeter of a regular pentagon with 6cm sides?

    ____

12. $\frac{1}{4}$ of 320 = ________

13. Which set has two equivalent fractions?

14. 19:50 = ________________ am ☐ pm ☐

15. $\frac{3}{4}$ = ________%

Problem-solving

Week 35

Mr Smith weighs 79kg. He has six children with the following weights: Koa 33kg, Kate 31kg, Karl 25kg, Kara 24kg, Kyle 23kg, and Keira 21kg.

Mr Smith sits on one side of a seesaw.

Let's find out which three children will balance the seesaw if they sit together on the other side.

Read the question again. **Think** about the information. Underline the important words.

Tick the strategy you will use to work out the answer:

- estimate and check ☐
- look for patterns ☐
- draw a diagram or picture ☐
- construct a table or graph ☐
- use materials ☐
- act it out ☐
- work backwards ☐
- something else. ☐

Solve it:

Reflect on the question and answer.

Check it. Circle another strategy on the list to work out the answer.

Show it:

Friday Review

1. 8097 ÷ 1000 = ______
2. 32 ÷ 8 = ______ ÷ 4 = 4
3. $\frac{1}{5}$ of 1000 = ______
4. The factors of 24 are: 1, ______, ______, ______, ______, ______, 12, 24
5. 99, 90, 81, ______, ______, 54
6. 90 ☐ 9 = 10
7. 998 + 8 = ______
8. 2⟌8008 = ______
9. $\frac{4}{16} = \frac{\square}{4}$
10. 0.6 × 5 = ______
11. 29 × 18 = ______
    30 × 18 = 540
    31 × 18 = ______
12. Rotate 90° anticlockwise.

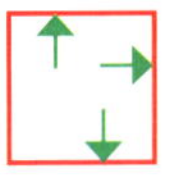

13. Write the 24-hour time for midnight.
    ______
14. Colour the sides parallel to the pink side.

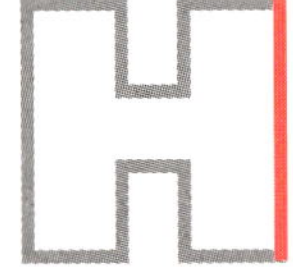

15. Balance the scales.

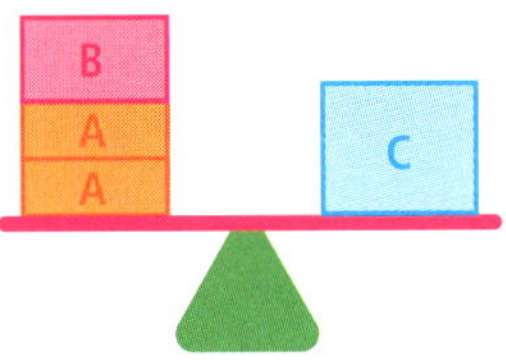

B = 40kg
A = 100kg
C = ______kg

16. Order the swimming times (shown in seconds) from fastest to slowest.

55.2 55.0
55.02 54.96

______ ______
______ ______

17. Complete the times with 15 minute intervals.

Departures
2:09 am
______ am
______ am

18. A rectangular yard has an area of $40m^2$ and a perimeter of 26m What is its:
    (a) length? ______
    (b) width? ______

N  A  M  Sp St  P 

Monday

1. $9.009 \times 1000 =$ ______

2. Chef paid $40 for oranges. How many kilograms did Chef buy?

______

3. $\frac{18}{100} + \frac{1}{10} = \frac{\square}{100} = 0.$ ______

4. $3\frac{2}{8} =$ ______ (improper fraction)

5. $\frac{1}{4} =$ ______ %

6. $25 \times 24 = 100 \times$ ______

7. $375 + 375 =$ ______

8. $4\overline{)2080} =$ ______

9. This is a ______.

10. Round 16.08 to the nearest tenth.

______

11. $\frac{1}{2} > \frac{1}{20}$ true ☐ false ☐

12. Rotate 90° clockwise.

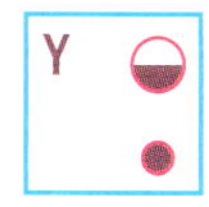

13. What is the value of the 4 in 2.4?

______

You are in charge of the class budget for the end-of-year party!

Class budget	
cash raised	$300
Expenses	
decorations	$30
plates and cutlery	$40
cleaning	$30
food	$150
drinks	$20

14. Total expenses = ______

15. Cash left = ______

Tuesday

Week 36

1. $4.09 \div 10 =$ ______

2. Write the next four multiples of 8.

136, ______, ______, ______, ______

3. Together, Jackson and Madison ate 27 chocolate-coated pineapple pieces. If Jackson ate twice as many as Madison, how many pieces did:

(a) Jackson eat? ______

(b) Madison eat? ______

4. Circle the vertex.

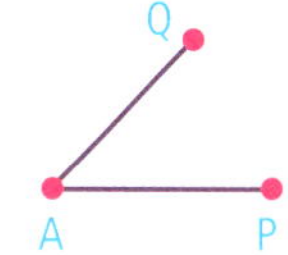

5. $2400 \div 5 =$ ______

6. Write $\frac{6}{8}$ in its simplest form. ______

7. An acute angle is less than ______.

8. What is the probability of you predicting the outcome of a coin flip correctly?

$\frac{1}{10}$ ☐ $\frac{1}{2}$ ☐ 1 ☐

9. Draw the lines of symmetry.

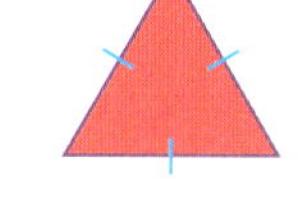

10. 23:00 = ______ am ☐ pm ☐

11. Arrange 8, 7, 9, and 0 to make the smallest number.

______

12. 1.01km = 1010m

1.02km = ______ m

13. Colour $\frac{2}{5}$ in blue.

14. Colour $\frac{2}{10}$ in green.

15. What fraction is uncoloured? ______

Week 36

Wednesday

1. 0.079 × 1000 = ________

2. Complete the compass rose.

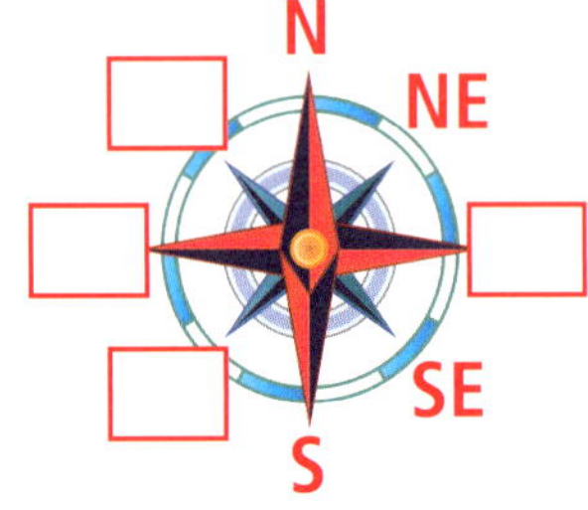

3.

15	30	45		
90				150

4. 255 000 + 75 000 = ________

5. 80 ☐ 80 = 0

6. 1.06 – 0.1 = ________

7. This is a ________.

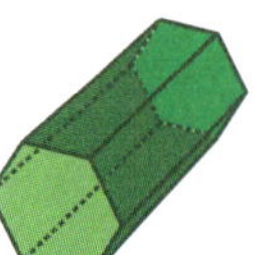

8. 1ha = ________ m^2

9. 2 × 38 × 50 = ________

10. Change $3\frac{3}{5}$ to an improper fraction.

____

11. What is the volume? ________

(1cm³ cubes)

12. 75% = $\frac{75}{100}$ = $\frac{3}{\square}$

13. 800 000 – 25 000 = ________

14. Flip, slide, or rotate the tetrominoes from A to fit into B.

A

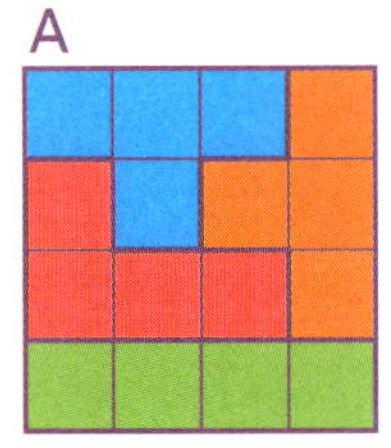

B

15. Angles A and ________ are acute.

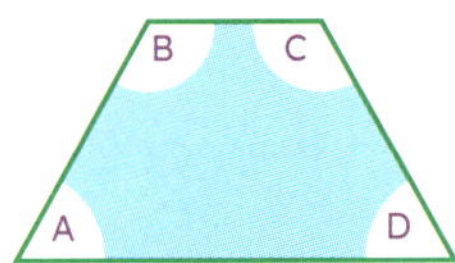

Thursday

1. 6.4 ÷ 100 = ________

2. Match each decimal with its sum.

0.46	$4 + \frac{6}{10}$
4.6	$\frac{4}{100} + \frac{6}{1000}$
0.046	$\frac{4}{10} + \frac{6}{100}$

3. 720 = ________ tens

4. $8 - \frac{1}{3}$ = ________

5. 3000 + 8000 = ________

6. Circle the numbers that are multiples of both 3 and 6.

12 15 33 24 32 42

7. 440 + 760 = ________

8. The probability of choosing a yellow is:

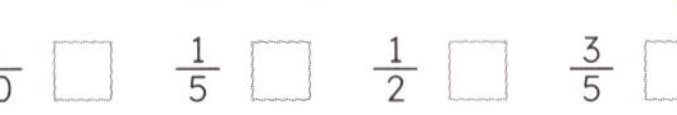

$\frac{1}{10}$ ☐ $\frac{1}{5}$ ☐ $\frac{1}{2}$ ☐ $\frac{3}{5}$ ☐

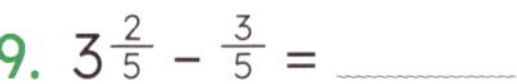

9. $3\frac{2}{5} - \frac{3}{5}$ = ________

10. How many $20 notes make $500?

____

11. 99 × 9 = ________

12. You paid $200 000 for this painting and sold it at 10% profit.

Profit = $________

ART SALE

13. 28 × 25 = ________ × 100

14. How many dried-out markers were found altogether?

____

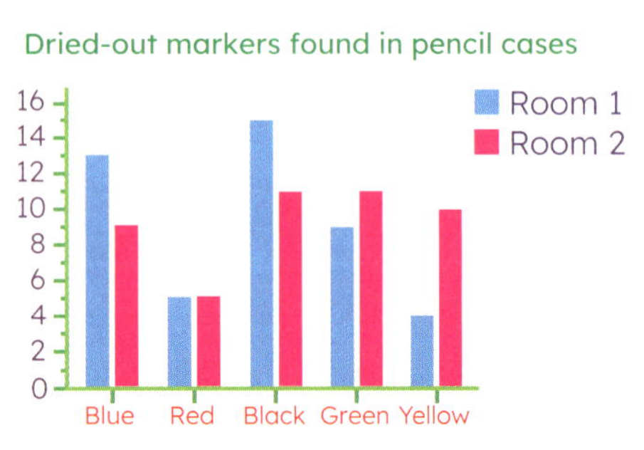

15. 1.04 – 0.1 = ________

Problem-solving

A ticket to the movie theatre costs $26.49 per adult and $19.99 per child. Groups larger than 10 people receive a 50% discount on all tickets.

Let's find out how much it would cost to buy a group of 10 adults and 10 children tickets.

Read the question again. **Think** about the information. Underline the important words.

Tick the strategy you will use to work out the answer:

- estimate and check ☐
- look for patterns ☐
- draw a diagram or picture ☐
- construct a table or graph ☐
- use materials ☐
- act it out ☐
- work backwards ☐
- something else. ☐

Solve it:

Reflect on the question and answer.

Check it. Circle another strategy on the list to work out the answer.

Show it:

Friday Review

Week 36

1. 1930 = ________ tens

2. This painting was $9000. It is now 50% off.

New price = ________

3. 0.92 + 0.1 = ________

4. $4\overline{)2040}$ = ________

5. Change $4\frac{3}{8}$ to an improper fraction. ________

6. Rotate 180°.

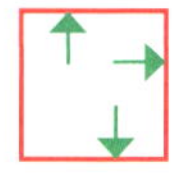 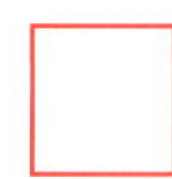

7. 2.07 × 10 000 = ________

8. In Thursday's graph, how many more green markers were dried-out in Room 2 than in Room 1? ________

9. $5 - \frac{4}{6}$ = ________

10. If you had 20 lots of 50c coins and 30 lots of 20c coins, what is the total amount? ________

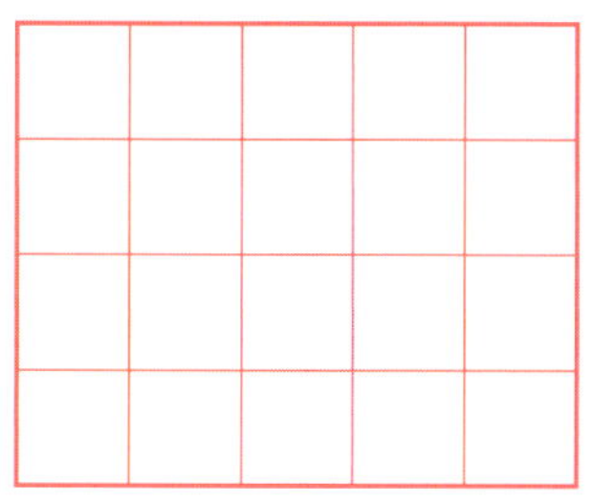

11. Colour $\frac{2}{5}$ in red.

12. Colour $\frac{1}{5}$ in blue.

13. What fraction is uncoloured? ________

14. 2ha = ________ m^2

15. 

= ________

am ☐ pm ☐

16. 9km 90m = ________ m

17. Draw ↑ to show the lightest side.

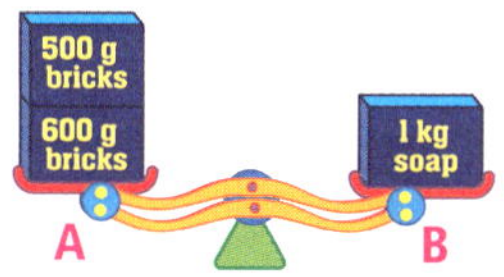

18. Using Thursday's marble bag, what is the probability of choosing a purple marble?

$\frac{1}{10}$ ☐ $\frac{1}{5}$ ☐
$\frac{1}{2}$ ☐ $\frac{3}{5}$ ☐

N A M    

Maths Facts

Number

Numbers 1–200

1	2	3	4	5	6	7	8	9	10	11	12	13	14	15	16	17	18	19	20
21	22	23	24	25	26	27	28	29	30	31	32	33	34	35	36	37	38	39	40
41	42	43	44	45	46	47	48	49	50	51	52	53	54	55	56	57	58	59	60
61	62	63	64	65	66	67	68	69	70	71	72	73	74	75	76	77	78	79	80
81	82	83	84	85	86	87	88	89	90	91	92	93	94	95	96	97	98	99	100
101	102	103	104	105	106	107	108	109	110	111	112	113	114	115	116	117	118	119	120
121	122	123	124	125	126	127	128	129	130	131	132	133	134	135	136	137	138	139	140
141	142	143	144	145	146	147	148	149	150	151	152	153	154	155	156	157	158	159	160
161	162	163	164	165	166	167	168	169	170	171	172	173	174	175	176	177	178	179	180
181	182	183	184	185	186	187	188	189	190	191	192	193	194	195	196	197	198	199	200

Suggested activity

Use a counter and a die or spinner.
Play against a partner. First to 200 wins!

What numbers make one hundred?

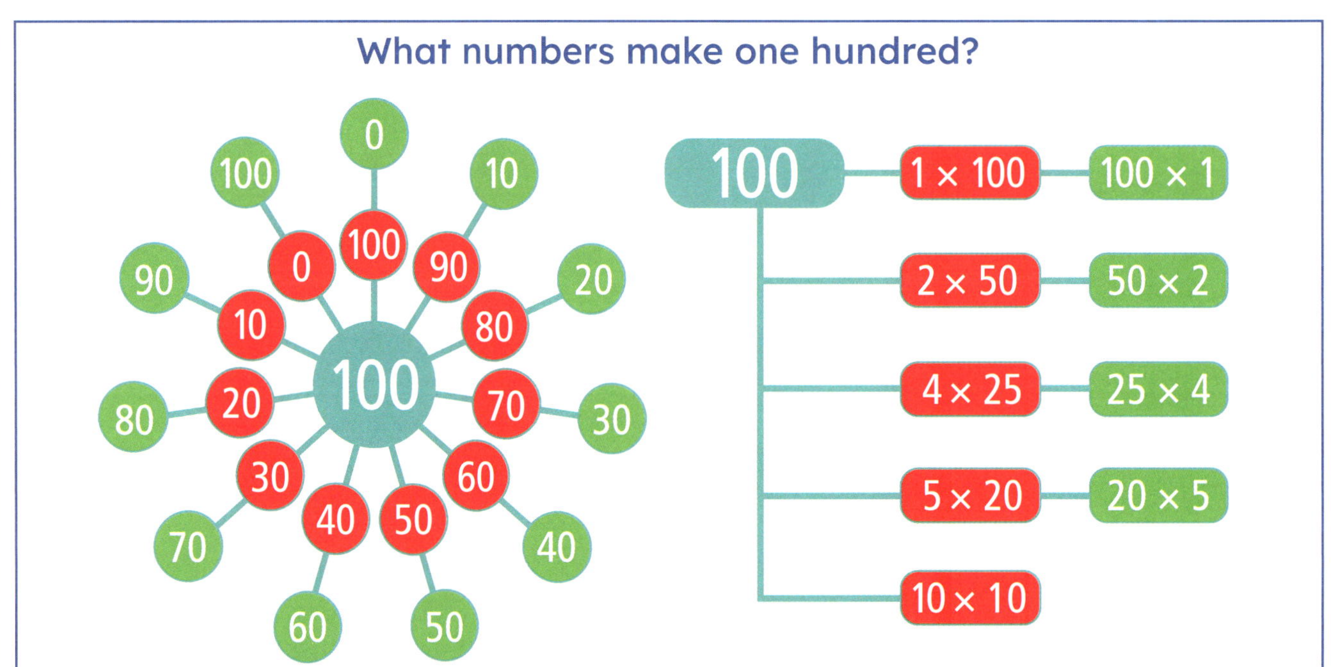

Maths Facts

Number

Multiplication and division facts

× 2	2 ×	÷ 2	quotient 2
1 × 2 = 2	2 × 1 = 2	2 ÷ 2 = 1	2 ÷ 1 = 2
2 × 2 = 4		4 ÷ 2 = 2	
3 × 2 = 6	2 × 3 = 6	6 ÷ 2 = 3	6 ÷ 3 = 2
4 × 2 = 8	2 × 4 = 8	8 ÷ 2 = 4	8 ÷ 4 = 2
5 × 2 = 10	2 × 5 = 10	10 ÷ 2 = 5	10 ÷ 5 = 2
6 × 2 = 12	2 × 6 = 12	12 ÷ 2 = 6	12 ÷ 6 = 2
7 × 2 = 14	2 × 7 = 14	14 ÷ 2 = 7	14 ÷ 7 = 2
8 × 2 = 16	2 × 8 = 16	16 ÷ 2 = 8	16 ÷ 8 = 2
9 × 2 = 18	2 × 9 = 18	18 ÷ 2 = 9	18 ÷ 9 = 2
10 × 2 = 20	2 × 10 = 20	20 ÷ 2 = 10	20 ÷ 10 = 2

× 3	3 ×	÷ 3	quotient 3
1 × 3 = 3	3 × 1 = 3	3 ÷ 3 = 1	3 ÷ 1 = 3
2 × 3 = 6	3 × 2 = 6	6 ÷ 3 = 2	6 ÷ 2 = 3
3 × 3 = 9		9 ÷ 3 = 3	
4 × 3 = 12	3 × 4 = 12	12 ÷ 3 = 4	12 ÷ 4 = 3
5 × 3 = 15	3 × 5 = 15	15 ÷ 3 = 5	15 ÷ 5 = 3
6 × 3 = 18	3 × 6 = 18	18 ÷ 3 = 6	18 ÷ 6 = 3
7 × 3 = 21	3 × 7 = 21	21 ÷ 3 = 7	21 ÷ 7 = 3
8 × 3 = 24	3 × 8 = 24	24 ÷ 3 = 8	24 ÷ 8 = 3
9 × 3 = 27	3 × 9 = 27	27 ÷ 3 = 9	27 ÷ 9 = 3
10 × 3 = 30	3 × 10 = 30	30 ÷ 3 = 10	30 ÷ 10 = 3

× 4	4 ×	÷ 4	quotient 4
1 × 4 = 4	4 × 1 = 4	4 ÷ 4 = 1	4 ÷ 1 = 4
2 × 4 = 8	4 × 2 = 8	8 ÷ 4 = 2	8 ÷ 2 = 4
3 × 4 = 12	4 × 3 = 12	12 ÷ 4 = 3	12 ÷ 3 = 4
4 × 4 = 16		16 ÷ 4 = 4	
5 × 4 = 20	4 × 5 = 20	20 ÷ 4 = 5	20 ÷ 5 = 4
6 × 4 = 24	4 × 6 = 24	24 ÷ 4 = 6	24 ÷ 6 = 4
7 × 4 = 28	4 × 7 = 28	28 ÷ 4 = 7	28 ÷ 7 = 4
8 × 4 = 32	4 × 8 = 32	32 ÷ 4 = 8	32 ÷ 8 = 4
9 × 4 = 36	4 × 9 = 36	36 ÷ 4 = 9	36 ÷ 9 = 4
10 × 4 = 40	4 × 10 = 40	40 ÷ 4 = 10	40 ÷ 10 = 4

× 5	5 ×	÷ 5	quotient 5
1 × 5 = 5	5 × 1 = 5	5 ÷ 5 = 1	5 ÷ 1 = 5
2 × 5 = 10	5 × 2 = 10	10 ÷ 5 = 2	10 ÷ 2 = 5
3 × 5 = 15	5 × 3 = 15	15 ÷ 5 = 3	15 ÷ 3 = 5
4 × 5 = 20	5 × 4 = 20	20 ÷ 5 = 4	20 ÷ 4 = 5
5 × 5 = 25		25 ÷ 5 = 5	
6 × 5 = 30	5 × 6 = 30	30 ÷ 5 = 6	30 ÷ 6 = 5
7 × 5 = 35	5 × 7 = 35	35 ÷ 5 = 7	35 ÷ 7 = 5
8 × 5 = 40	5 × 8 = 40	40 ÷ 5 = 8	40 ÷ 8 = 5
9 × 5 = 45	5 × 9 = 45	45 ÷ 5 = 9	45 ÷ 9 = 5
10 × 5 = 50	5 × 10 = 50	50 ÷ 5 = 10	50 ÷ 10 = 5

× 6	6 ×	÷ 6	quotient 6
1 × 6 = 6	6 × 1 = 6	6 ÷ 6 = 1	6 ÷ 1 = 6
2 × 6 = 12	6 × 2 = 12	12 ÷ 6 = 2	12 ÷ 2 = 6
3 × 6 = 18	6 × 3 = 18	18 ÷ 6 = 3	18 ÷ 3 = 6
4 × 6 = 24	6 × 4 = 24	24 ÷ 6 = 4	24 ÷ 4 = 6
5 × 6 = 30	6 × 5 = 30	30 ÷ 6 = 5	30 ÷ 5 = 6
6 × 6 = 36		36 ÷ 6 = 6	
7 × 6 = 42	6 × 7 = 42	42 ÷ 6 = 7	42 ÷ 7 = 6
8 × 6 = 48	6 × 8 = 48	48 ÷ 6 = 8	48 ÷ 8 = 6
9 × 6 = 54	6 × 9 = 54	54 ÷ 6 = 9	54 ÷ 9 = 6
10 × 6 = 60	6 × 10 = 60	60 ÷ 6 = 10	60 ÷ 10 = 6

× 7	7 ×	÷ 7	quotient 7
1 × 7 = 7	7 × 1 = 7	7 ÷ 7 = 1	7 ÷ 1 = 7
2 × 7 = 14	7 × 2 = 14	14 ÷ 7 = 2	14 ÷ 2 = 7
3 × 7 = 21	7 × 3 = 21	21 ÷ 7 = 3	21 ÷ 3 = 7
4 × 7 = 28	7 × 4 = 28	28 ÷ 7 = 4	28 ÷ 4 = 7
5 × 7 = 35	7 × 5 = 35	35 ÷ 7 = 5	35 ÷ 5 = 7
6 × 7 = 42	7 × 6 = 42	42 ÷ 7 = 6	42 ÷ 6 = 7
7 × 7 = 49		49 ÷ 7 = 7	
8 × 7 = 56	7 × 8 = 56	56 ÷ 7 = 8	56 ÷ 8 = 7
9 × 7 = 63	7 × 9 = 63	63 ÷ 7 = 9	63 ÷ 9 = 7
10 × 7 = 70	7 × 10 = 70	70 ÷ 7 = 10	70 ÷ 10 = 7

× 8	8 ×	÷ 8	quotient 8
1 × 8 = 8	8 × 1 = 8	8 ÷ 8 = 1	8 ÷ 1 = 8
2 × 8 = 16	8 × 2 = 16	16 ÷ 8 = 2	16 ÷ 2 = 8
3 × 8 = 24	8 × 3 = 24	24 ÷ 8 = 3	24 ÷ 3 = 8
4 × 8 = 32	8 × 4 = 32	32 ÷ 8 = 4	32 ÷ 4 = 8
5 × 8 = 40	8 × 5 = 40	40 ÷ 8 = 5	40 ÷ 5 = 8
6 × 8 = 48	8 × 6 = 48	48 ÷ 8 = 6	48 ÷ 6 = 8
7 × 8 = 56	8 × 7 = 56	56 ÷ 8 = 7	56 ÷ 7 = 8
8 × 8 = 64		64 ÷ 8 = 8	
9 × 8 = 72	8 × 9 = 72	72 ÷ 8 = 9	72 ÷ 9 = 8
10 × 8 = 80	8 × 10 = 80	80 ÷ 8 = 10	80 ÷ 10 = 8

× 9	9 ×	÷ 9	quotient 9
1 × 9 = 9	9 × 1 = 9	9 ÷ 9 = 1	9 ÷ 1 = 9
2 × 9 = 18	9 × 2 = 18	18 ÷ 9 = 2	18 ÷ 2 = 9
3 × 9 = 27	9 × 3 = 27	27 ÷ 9 = 3	27 ÷ 3 = 9
4 × 9 = 36	9 × 4 = 36	36 ÷ 9 = 4	36 ÷ 4 = 9
5 × 9 = 45	9 × 5 = 45	45 ÷ 9 = 5	45 ÷ 5 = 9
6 × 9 = 54	9 × 6 = 54	54 ÷ 9 = 6	54 ÷ 6 = 9
7 × 9 = 63	9 × 7 = 63	63 ÷ 9 = 7	63 ÷ 7 = 9
8 × 9 = 72	9 × 8 = 72	72 ÷ 9 = 8	72 ÷ 8 = 9
9 × 9 = 81		81 ÷ 9 = 9	
10 × 9 = 90	9 × 10 = 90	90 ÷ 9 = 10	90 ÷ 10 = 9

Maths Facts

Number

Multiplication and division facts

× 10	10 ×	÷ 10	quotient 10
1 × 10 = 10	10 × 1 = 10	10 ÷ 10 = 1	10 ÷ 1 = 10
2 × 10 = 20	10 × 2 = 20	20 ÷ 10 = 2	20 ÷ 2 = 10
3 × 10 = 30	10 × 3 = 30	30 ÷ 10 = 3	30 ÷ 3 = 10
4 × 10 = 40	10 × 4 = 40	40 ÷ 10 = 4	40 ÷ 4 = 10
5 × 10 = 50	10 × 5 = 50	50 ÷ 10 = 5	50 ÷ 5 = 10
6 × 10 = 60	10 × 6 = 60	60 ÷ 10 = 6	60 ÷ 6 = 10
7 × 10 = 70	10 × 7 = 70	70 ÷ 10 = 7	70 ÷ 7 = 10
8 × 10 = 80	10 × 8 = 80	80 ÷ 10 = 8	80 ÷ 8 = 10
9 × 10 = 90	10 × 9 = 90	90 ÷ 10 = 9	90 ÷ 9 = 10
10 × 10 = 100		100 ÷ 10 = 10	

Odd and even numbers

Odd numbers up to 20:
1, 3, 5, 7, 9, 11, 13, 15, 17, 19

Even numbers up to 20:
2, 4, 6, 8, 10, 12, 14, 16, 18, 20

Place value

1 to 9	one digit	ones
10 to 99	two digits	tens and ones
100 to 999	three digits	hundreds, tens, and ones
1000 to 9999	four digits	thousands, hundreds, tens, and ones
10 000 to 99 999	five digits	ten thousands, thousands, hundreds, tens, and ones
100 000 to 999 999	six digits	hundred thousands, ten thousands, thousands, hundreds, tens, and ones
1 000 000 to 9 999 999	seven digits	millions, hundred thousands, ten thousands, thousands, hundreds, tens, and ones

Fractions

1 whole

$\frac{1}{2}$ $\frac{1}{2}$

$\frac{1}{3}$ $\frac{1}{3}$ $\frac{1}{3}$

$\frac{1}{4}$ $\frac{1}{4}$ $\frac{1}{4}$ $\frac{1}{4}$

$\frac{1}{5}$ $\frac{1}{5}$ $\frac{1}{5}$ $\frac{1}{5}$ $\frac{1}{5}$

$\frac{1}{6}$ $\frac{1}{6}$ $\frac{1}{6}$ $\frac{1}{6}$ $\frac{1}{6}$ $\frac{1}{6}$

$\frac{1}{8}$ $\frac{1}{8}$ $\frac{1}{8}$ $\frac{1}{8}$ $\frac{1}{8}$ $\frac{1}{8}$ $\frac{1}{8}$ $\frac{1}{8}$

$\frac{1}{10}$ $\frac{1}{10}$ $\frac{1}{10}$ $\frac{1}{10}$ $\frac{1}{10}$ $\frac{1}{10}$ $\frac{1}{10}$ $\frac{1}{10}$ $\frac{1}{10}$ $\frac{1}{10}$

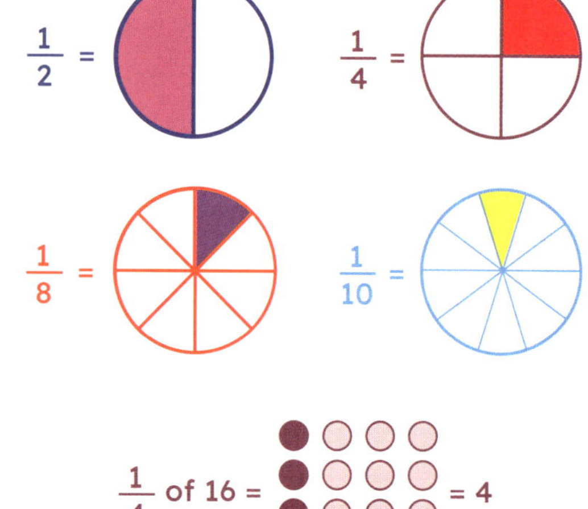

Maths Facts

Cross Strand Number and Measurement

What coins make one dollar?

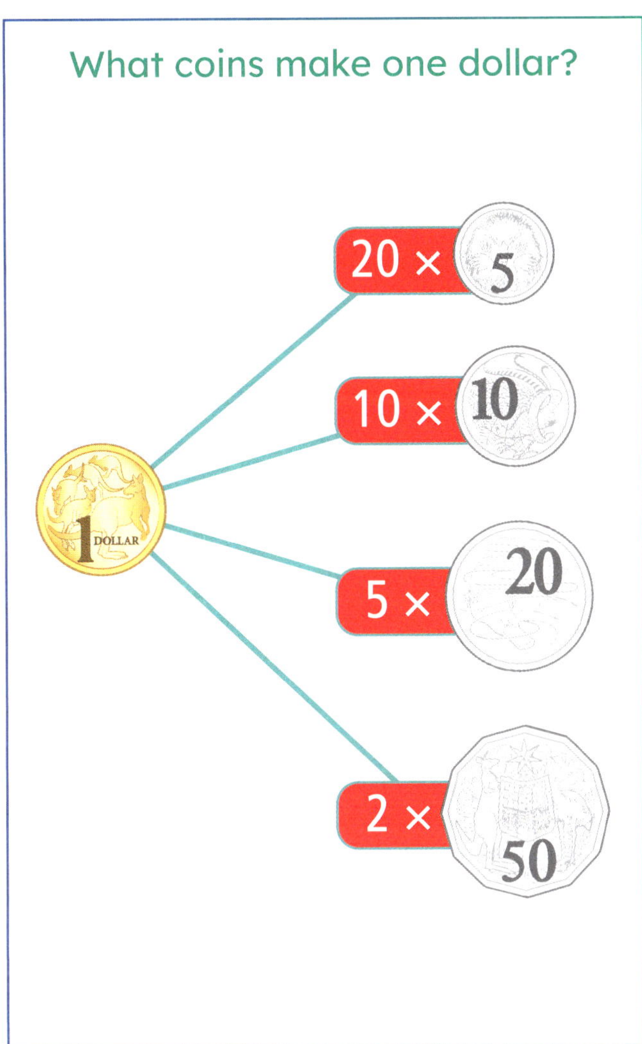

What coins make two dollars?

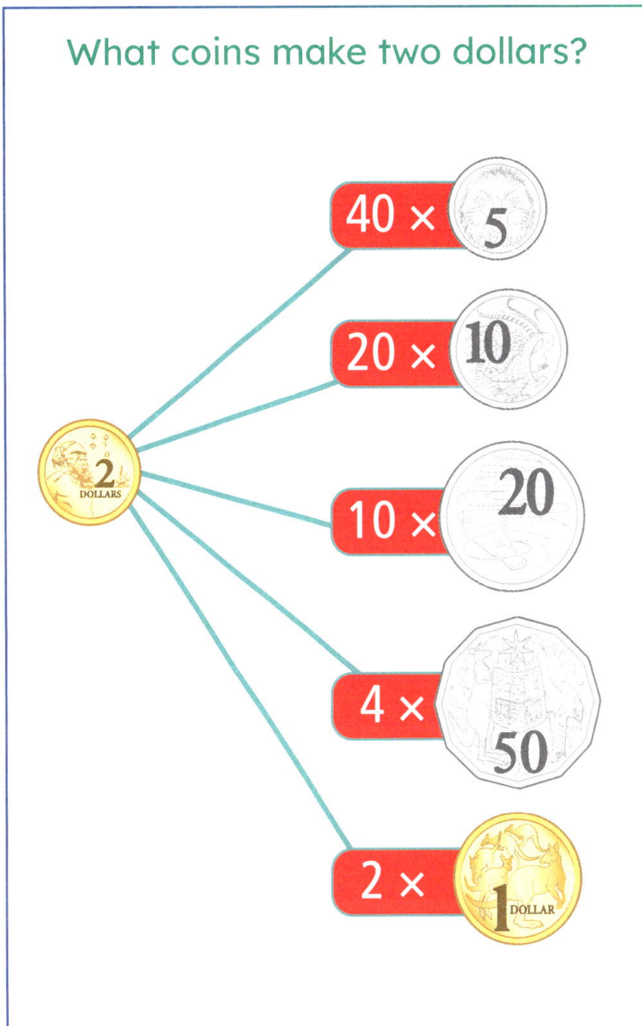

Algebra

Patterns – examples

1, 2, 3, 4, 5	Rule: + 1	50, 45, 40, 35, 30	Rule: – 5
19, 17, 15, 13, 11	Rule: – 2	32, 42, 52, 62, 72	Rule: + 10
3, 6, 9, 12, 15	Rule: + 3	99, 89, 79, 69, 59	Rule: – 10
5, 9, 13, 17, 21	Rule: + 4	1, 2, 4, 7, 11	Rule: + 1, + 2, + 3, + 4

Inverse operations

An inverse operation is an operation that reverses the effect of another operation.

Addition and subtraction are inverse operations:

5 + 2 = 7

7 – 5 = 2

Multiplication and division are inverse operations:

5 × 2 = 10

10 ÷ 5 = 2

Maths Facts

Measurement

Capacity

Unit	Abbreviation
millilitre	mL
litre	L

1000mL = 1L

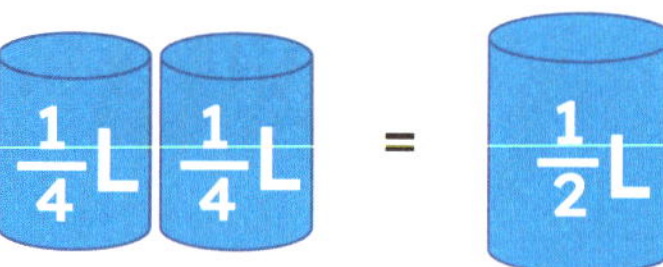

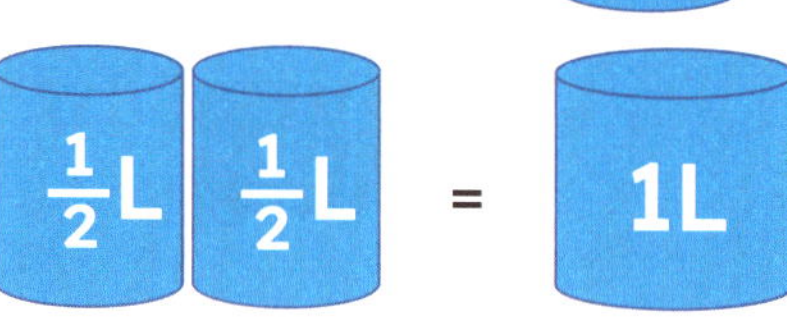

Mass

Unit	Abbreviation
gram	g
kilogram	kg
tonne	t

1000g = 1kg

1000kg = 1t

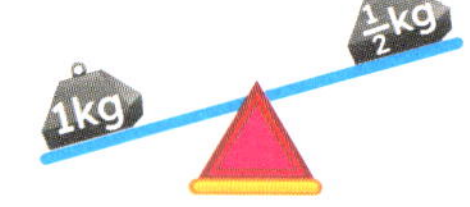

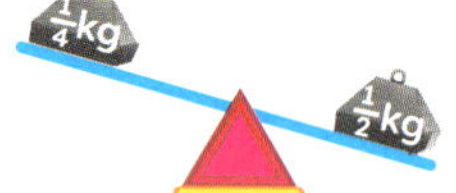

Angles

acute angle

An acute angle is less than 90°.

right angle

A right angle is 90°.

obtuse angle

An obtuse angle is between 90° and 180°.

reflex angle

A reflex angle is between 180° and 360°.

Time

am
The time between midnight and midday is am time.

pm
The time between midday and midnight is pm time.

9:00 am or 9:00 pm

The **24-hour clock** runs from 0000 (midnight) to 2359.

0900 or 9:00 am

2100 or 9:00 pm

Length

Unit	Abbreviation
millimetre	mm
centimetre	cm
metre	m
kilometre	km

10mm = 1cm | 100cm = 1m

1000mm = 1m | 1000m = 1km

Area

The area of a rectangle can be found by applying the formula:

area = length × width

Examples

Area = L × W
Area = 3cm × 2cm
Area = $6cm^2$

Area = L × W
Area = 6m × 1m
Area = $6m^2$

1 hectare = $10\,000m^2$

Maths Facts

Cross Strand Measurement and Space

Parallel lines

Parallel lines do not meet.

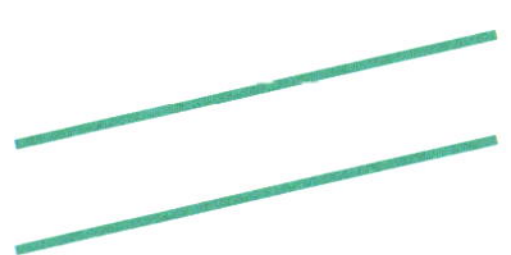

Examples

railway tracks

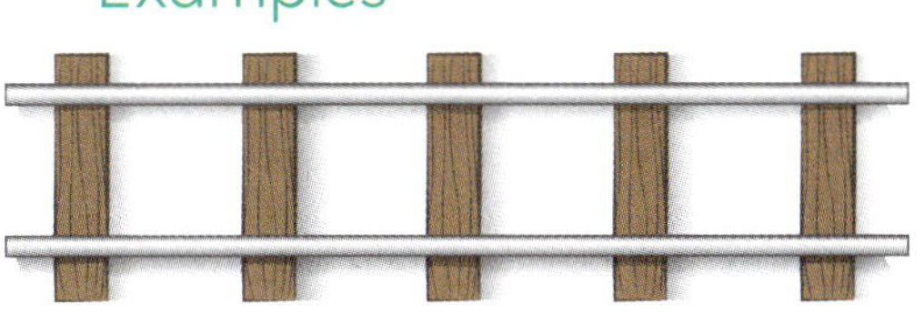

streets

Orientation

horizontal line

vertical line

Space

Directions

Compass

N means **north**

E means **east**

S means **south**

W means **west**

Compass directions

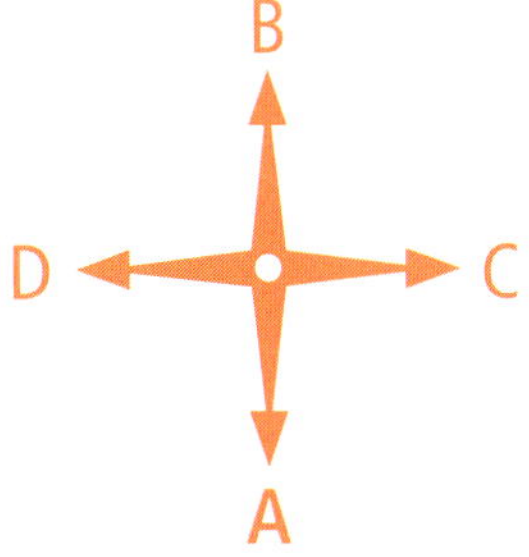

B is **north** of A

A is **south** of B

C is **east** of D

D is **west** of C

1 turn (quarter turn)
clockwise

1 turn (quarter turn)
anticlockwise

2 turns (half turn)
clockwise
or **anticlockwise**

Maths Facts

Space

2D shapes – quadrilaterals

A quadrilateral is a shape with 4 sides and 4 angles.
The total of the angles is 360°.

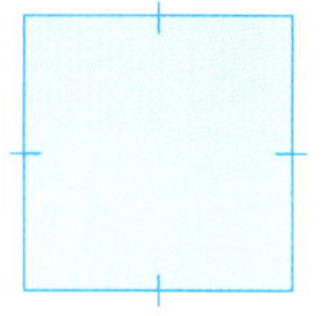

square
4 sides the same length.
4 right angles.

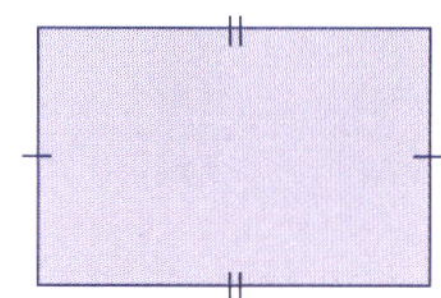

rectangle
2 pairs of sides the same length.
4 right angles.

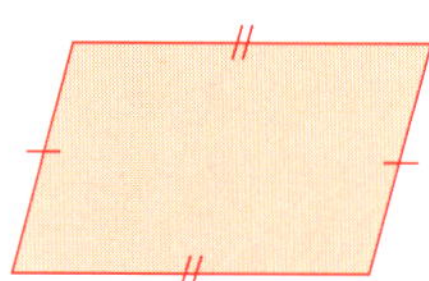

parallelogram
2 pairs of sides the same length. Opposite angles are equal.

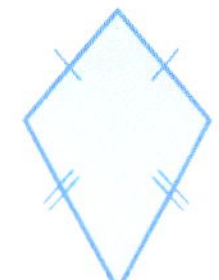

kite
No parallel sides.

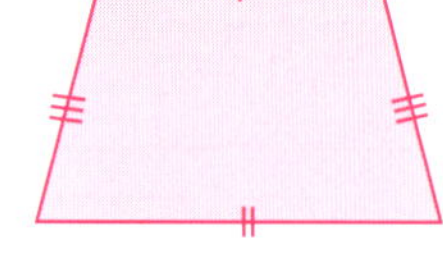

trapezium
1 pair of parallel sides.

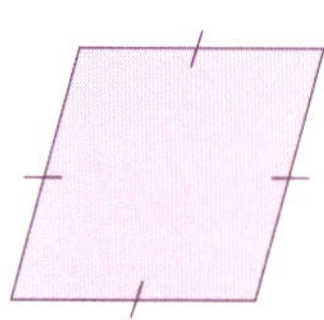

rhombus
4 sides the same length.
Opposite angles are equal.

Suggested activities

1. Trace or cut out. Reflect (flip) the shapes.
2. Measure the angles in the rhombus and the square. How are they different or similar?
3. Draw patterns using 2 or more of the quadrilaterals.

Other 2D shapes

circle

pentagon
5 sides
5 corners

hexagon
6 sides
6 corners

octagon
8 sides
8 corners

semicircle

irregular pentagon

irregular hexagon

irregular octagon

2D shapes – triangles

A triangle is a shape with 3 sides and 3 angles. The total of the angles is 180°.

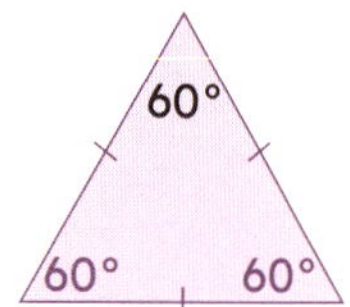

equilateral triangle
3 sides the same length.
3 angles the same size.

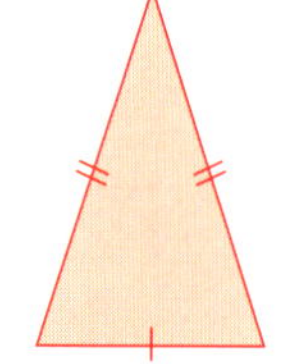

isosceles triangle
2 sides the same length.
2 angles the same size.

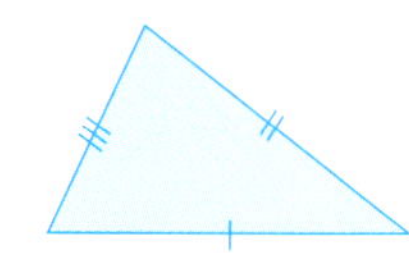

scalene triangle
No equal sides or angles.

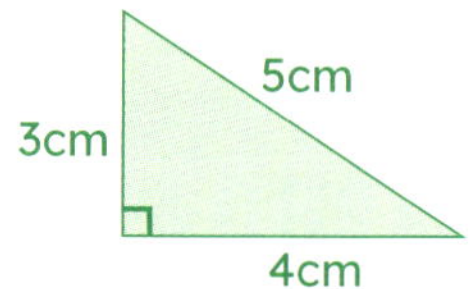

right-angled triangle
1 angle that is 90°.
Example:
a 3, 4, 5 triangle (scalene) makes a right angle.

Maths Facts

Space

3D objects

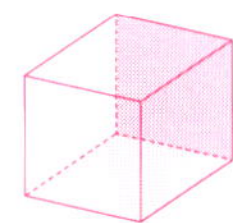

cube
6 faces
12 edges
8 vertices

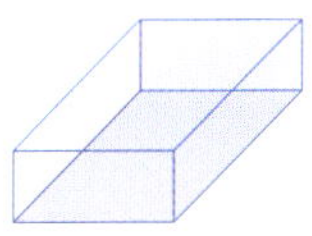

rectangular prism
6 faces
12 edges
8 vertices

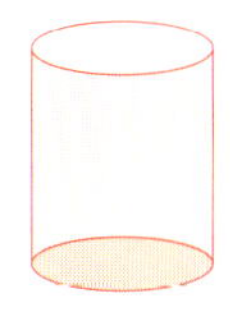

cylinder
2 flat surfaces
1 round surface

cone
1 flat surface
1 round surface
1 vertex

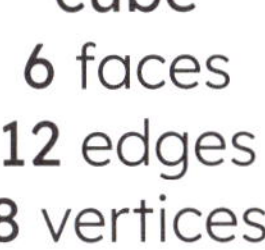

sphere
1 round surface

triangular prism
5 faces
9 edges
6 vertices

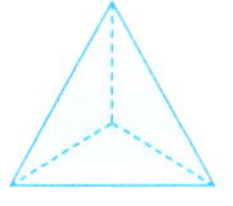

triangular pyramid
4 faces
6 edges
4 vertices

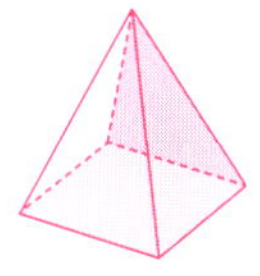

square pyramid
5 faces
8 edges
5 vertices

3D object nets

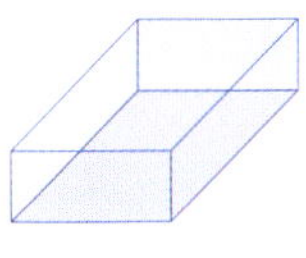

rectangular prism

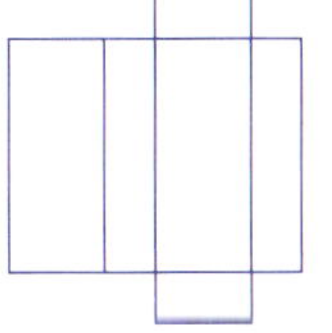

triangular prism

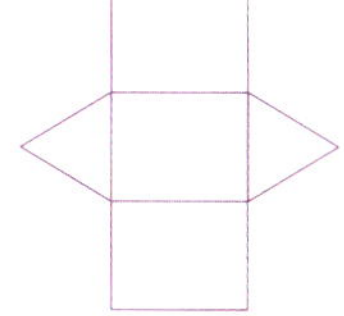

cylinder

triangular pyramid

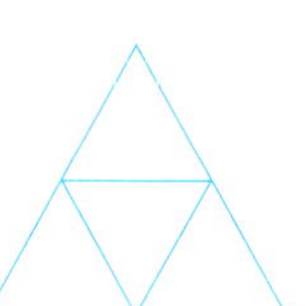

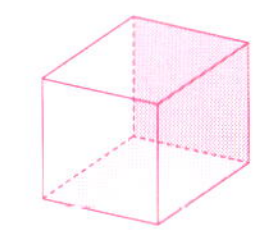

cube

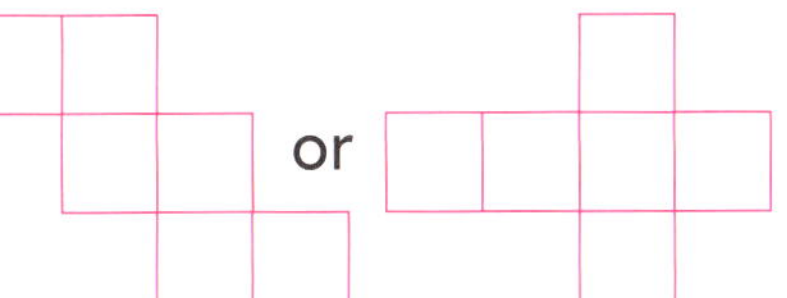

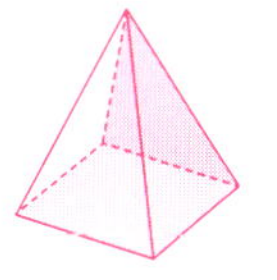

square pyramid

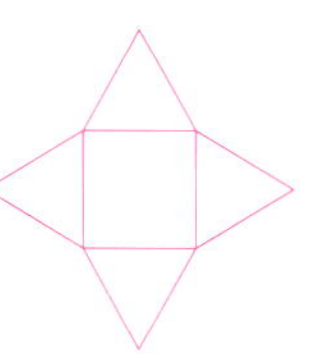

Symmetry

A square has 4 lines of symmetry.

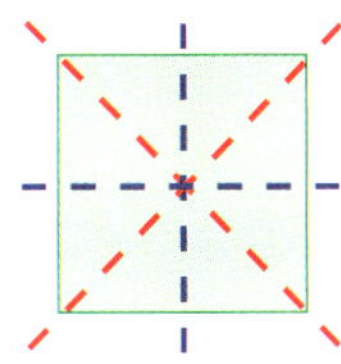

A rectangle has 2 lines of symmetry.

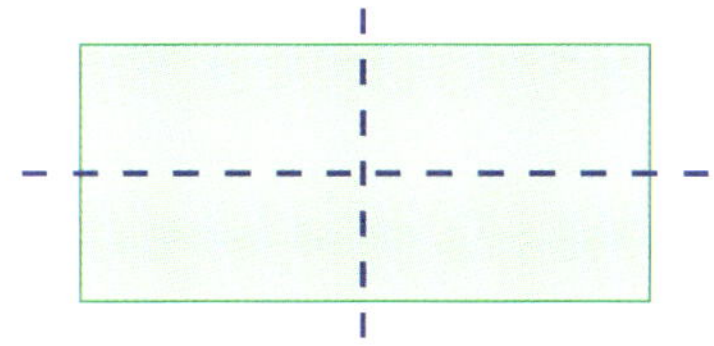

Maths Facts

Statistics

Picture graph

A graph in which data is represented by pictures. One picture could represent 1 unit or many.

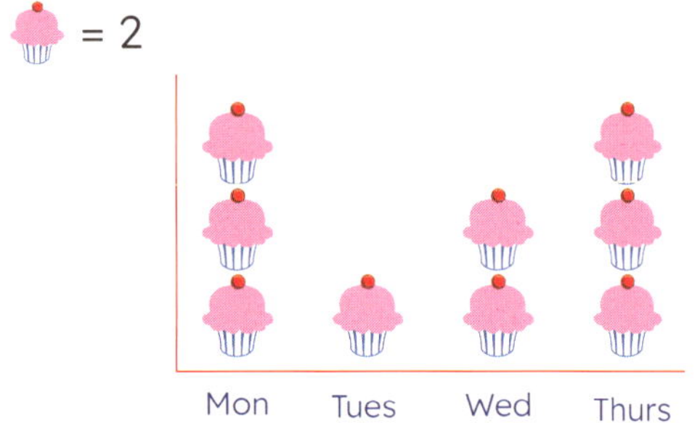

Bar graph / Column graph

A graph in which data is represented by bars. The height or length of the bars represents the units being measured.

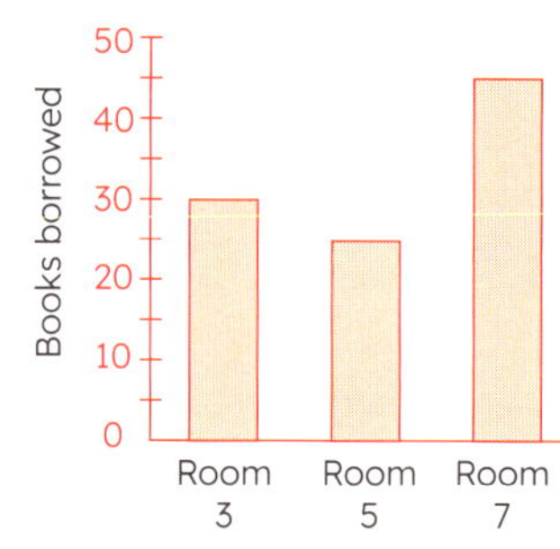

Tally

| = 1   𝍸 = 5   𝍸 𝍸 = 10   𝍸 𝍸 𝍸 𝍸 = 20

Probability

Chance

The likelihood of an event occurring.

Outcomes

Rolling a die – (1, 2, 3, 4, 5, 6)

Coin toss – (heads, tails)

An even number on a die – 3 in 6 or 1 in 2

Working Out Space